PATRICK MOORE'S
ARMCHAIR ASTRONOMY

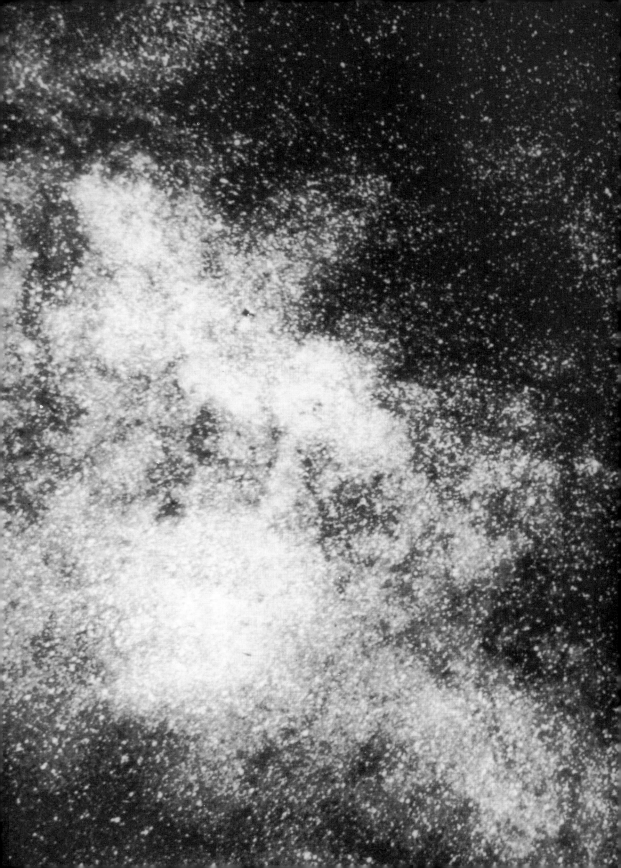

PATRICK MOORE'S ARMCHAIR ASTRONOMY

PSL

Patrick Stephens, Wellingborough

First published in 1984

British Library Cataloguing in Publication Data

Moore, Patrick
Patrick Moore's book of armchair astronomy.
1. Outer space—Miscellanea
I. Title
001.9'4'0999   Q8500

ISBN 0-85059-718-8

Patrick Stephens Limited is part of the
Thorsons Publishing Group.

Photoset in 10 on 11pt Garamond by MJL Typesetting, Hitchin, Herts. Printed in Great Britain on 115 gsm Fineblade coated cartridge, and bound, by The Garden City Press, Letchworth, Herts, for the publishers, Patrick Stephens Limited, Denington Estate, Wellingborough, Northants, NN8 2QD, England.

CONTENTS

FOREWORD

Most books on astronomy are written to a set, logical plan. This one is different. It has no plan at all.

What I have tried to do is to present a series of what may be termed 'snippets', with no particular theme, but ranging over the whole astronomical scene—from the Apollo moon-men to the Venerable Bede, from the inner Solar System out to the remotest quasar, from the magnificent work of modern astronomers to the clumsy forgeries of the Chevalier D'Angos. So if the night is cloudy, and you are waiting impatiently to open your dome and set to work, I hope you will feel like opening this book at random and seeing what you find. I have tried to avoid overlap, and also to keep mainly to what could be called 'way-out' topics. Perhaps you will find something here to interest or amuse you until the sky clears once more.

Patrick Moore, Selsey

April 1984

ACKNOWLEDGEMENTS

My grateful thanks are due to Paul Doherty for his splendid illustrations, to Robert Allen (who first suggested that I should write the book) and to Bruce Quarrie and Darryl Reach for all their help and encouragement.

P.M.

1 HOMESTAKE MINE

In the early part of 1982 I made a journey from my home in the Sussex village of Selsey. I went to Deadwood Gulch in South Dakota, where my main purpose was to go down a mile-deep gold-mine and look at a large tank of cleaning fluid.

This may sound somewhat eccentric, but there was a good reason for it, because I was going to what is certainly the world's oddest observatory. It has been set up to study the Sun, but as it is a mile down in Homestake Gold Mine it can never see the Sun or any other part of the sky. Once in the observatory rooms there is a mile of solid rock above you.

Deadwood was the country of the gunslingers of little more than a century ago: Wild Bill Hickok, Calamity Jane, Dr John F. ('Doc') Holliday and the rest. The gunslingers have gone, but the gold remains, and Homestake Mine is the largest in America. The observatory has been built at the end of a long tunnel, in two large caverns specially hollowed out by the gold-mine authorities. To get there you go down in the miners' cage (something of a rough ride) and then, putting on hard helmets and other pieces of safety equipment, make your way down the tunnel to the observatory—taking care not to touch the overhead cable, which carries enough electric current to fry you like an egg.

The 'telescope' is a tank holding 100,000 gallons of cleaning fluid. The chemical formula for cleaning fluid is C_2Cl_4. This means that it is made up of carbon and chlorine, and in this context it is the chlorine which matters.

Now let me explain the meaning of all this. The Sun sends out streams of particles called neutrinos, which are very difficult to study because they have no mass (or virtually none) and no electrical charge. They can pass unchecked through almost everything—right through the Earth, for instance—and you are being bombarded by neutrinos at this very moment, though I hasten to add that they are quite harmless.

Theorists are very anxious to know how many neutrinos the Sun is emitting, if only because only neutrinos can bring us up-to-date information from the 'power-house' at the solar core. The only way to catch the neutrinos is to make them interact with atoms of chlorine. This happens only occasionally, but when it does, the chlorine is changed into a different type of atom, argon-37, which is radioactive and can be detected.

Therefore, the cleaning fluid in the tank is left for a period of weeks and then checked. The amount of argon-37 shows how many neutrinos have scored direct hits on the chlorine atoms in the tank, and this in turn gives the number of neutrinos which the Sun is sending out. You have to go underground because

The entrance to Homestake Mine. The solar observatory is in a shaft a mile below ground level!

high-speed particles from space, known as cosmic rays, produce the same effect. Cosmic rays cannot penetrate through a mile of rock; only neutrinos can do so.

At present the experiments are giving unexpected results, and there are fewer solar neutrinos than had been anticipated. This may show that there is something wrong with our theories of the Sun. So Homestake Mine is a vital link in the chain—and certainly it is unlike any other observatory that I have had the honour to visit!

2   NEW MOON?

One new moon occurs every 29 ½ days; one full moon also occurs every 29 ½ days. Everyone knows the full moon, but how many people have seen a new moon? The surprising answer is: 'Not so many as might be imagined', because in the ordinary way the real new moon is invisible. When the Moon appears in the evening sky as a slender crescent, it is past 'new'.

As we know, the new moon occurs when the Moon's unlit hemisphere is turned towards us. The night side of the Moon does not shine at all. If the lining-up is exact, then there is a solar eclipse. On these occasions you will see the dark disk of the genuinely new Moon encroaching on the Sun, and if the eclipse is total the sight is magnificent. For a few fleeting minutes (never as long as eight) the Sun's atmosphere shines out from behind the Moon. But solar eclipses do not happen every month, because the lunar orbit is appreciably inclined, and at most new moons the Moon passes invisibly either above or below the Sun in the sky.

All sorts of myths and legends have grown up around the new moon. In particular, there are still people who believe that the phases of the Moon affect the weather. In fact there is no connection, and there is no conceivable reason why there should be, because when the Moon is new, it is not necessarily at its closest to us. The lunar orbit round the Earth (or, to be more precise, round the centre of gravity of the Earth-Moon system) is not circular, and so the distance ranges between 221,460 miles at closest approach (perigee) and 252,700 miles at furthest recession (apogee), but perigee and apogee may occur at any phase.

Associations between the lunar phase and plant life have been suggested: I remember a German female experimenter, L. Kolisko of Stuttgart, who carried out studies between 1926 and 1935 claiming, for instance, that tomatoes sown two days before full moon were stronger and juicier than those sown two days before new moon; she went on to state that while all full-moon tomato plants flourished, a considerable number of new-moon plants died. Finally, she added that the Easter full and new moons were of special significance.

Kolisko, of course, was an eccentric, and cannot be taken seriously; but even today there are many people who believe in a connection between the Moon's phase and mental health, so that unstable people are at their worst when the moon is full and at their best when it is new. Some years ago I put the question to 200 doctors, and found that while most of them were sceptical there were at least 30 who believed in some kind of effect, though they could not give any reason for it. In a further investigation I consulted Hansard, in which the speeches of Members of Parliament are recorded. I wanted to see whether the standard of intelligence was any better at new moon than when the Moon was full. However, I found that the overall standard was so consistently inane that I could draw no conclusions from it!

3 ASTRONOMY AND THE VENERABLE BEDE

Looking back over the list of British amateur astronomers, there are reasons for suggesting that one of the earliest of them—if not the earliest of all—was the Venerable Bede. Almost everybody has heard his name, though not so many know where he lived or what he did.

Bede was born in the year 673 at Monkwearmouth, in what is now north-east Durham. At that stage in history, of course, there was no single kingdom of England, and there were several independent kingdoms of which Northumbria was the most powerful. There had been wars between Northumbria and the 'middle kingdom' of Mercia, which had been ruled by King Penda, one of the last champions of heathenism in Britain. In 642 King Oswald of Northumbria had been defeated and killed by Penda's forces, but in 655, at the Battle of Winwæd, it was Northumbria's turn: the new king, Oswy, routed Penda's forces and killed Penda himself. By the time of Bede's birth the conversion of England to Christianity was almost complete, though Sussex held out for a few years longer.

When Oswy died, Egfrith ascended the throne of Northumbria, and for a while things were peaceful. Certainly Bede did not lead an adventurous life. He became a monk, and spent his days in a monastery at Jarrow, writing voluminously about all sorts of subjects. Evidently he had no liking for travel, and it is possible that he never even visited Hadrian's Wall, though he lived less than ten miles away from it.

Bede's greatest book was his *Ecclesiastical History of the English People*, completed in 731. In it there are several references to comets. Thus 'In the month of August 678, in the eighth year of Egfrith's reign, a star known as a comet appeared, which remained visible for three months, rising in the morning and emitting what seemed to be a tall column of bright flame.' Actually Bede may have the date wrong: according to Chinese sources there was a brilliant comet in September 676, which had a 3-degree tail and moved from Gemini into the region of the Great Bear before it faded away.

Then, in 729, Bede wrote that 'two comets appeared around the Sun, striking terror into all who saw them. One comet rose early and preceded the Sun, while the other followed the setting sun at evening. They appeared in January, and remained visible for about two weeks, bearing their fiery tails northward as though to set the welkin aflame.' Probably Bede himself saw this comet—it may well have been one rather than two, rising before the Sun in the morning and setting after it in the evening. Bede does not mention Halley's Comet, which appeared during his lifetime; but it was visible when Bede was only 11 years old, so that it is understandable that he did not record it.

The power of Northumbria declined during Bede's career. King Egfrith became over-bold; he led an army across the Forth, and was routed by the Picts, in 685, at the Battle of Nectansmere. But at least Bede missed the Vikings, who did not arrive upon the English scene until about 789.

Bede died in 731, busy with his writing almost to the end. His *History* is invaluable; it gives us our only reliable source of information about what happened in England for the first few centuries after the departure of the Romans. Note, too, that when the Jesuit moon-mapper Riccioli allotted names to lunar craters in 1651, Bede was commemorated by a rather large though obscure ring near the bright crater we now call Censorinus. For some reason or other the name was dropped by later astronomers; but I feel that it should be restored, if only because Bede was at least paying some attention to the sky when nobody else in England was likely to be doing so.

4 ASTEROIDS WITH ATMOSPHERES?

The asteroids, or minor planets, are junior members of the Sun's family. All the larger members of the swarm keep strictly to the region between the orbits of Mars and Jupiter (unless we except the strange Chiron, which moves much further out and spends most of its time between the orbits of Saturn and Uranus).

The first four asteroids to be discovered, between 1801 and 1807, were Ceres, Pallas, Juno and Vesta. Ceres, with a diameter of over 600 miles, is much the largest of all the asteroids, while Vesta is the only one ever visible to the naked eye. Needless to say, all are much too small to retain any trace of atmosphere. It is therefore rather interesting to look back at some very early ideas about them.

One of the first popular books on astronomy to be published was by James Ferguson. It was revised several times, and in the 1811 edition, revised by David Brewster, we find descriptions of the 'new planets'. I quote from page 126 of the revision:

'The planet Ceres was discovered at Palermo, in Sicily, on 1 January 1801, by M. Piazzi. . . . This new celestial body was then situated in Taurus, and was observed by Piazzi till 12 February, when a dangerous illness compelled him to discontinue his observations. It was, however, again discovered by Dr. Olbers, of Bremen, on 1 January 1802*, nearly in the place where it was expected from the calculations of Baron Zach. The nebula with which it was surrounded, gave it the appearance of a comet. . . . The planet Ceres is of a ruddy colour, and seems to be surrounded with a large dense atmosphere, and plainly exhibits a disk when examined with a magnifying power of about 200.' Brewster adds that William Herschel estimated the diameter to be 160 miles, while Johann Schröter gave an estimate of 1,624 miles!

Asteroid number 2, Pallas, was found a year later. To quote Brewster again: 'It is situated between the orbits of Mars and Jupiter, and is nearly of the same magnitude as Ceres, but of a less ruddy colour. It is surrounded with a nebulosity of almost the same extent. . . . The diameter of Pallas has not yet been determined with sufficient accuracy. Dr. Herschel makes it only 80 miles, while Schröter makes it no less than 2,099 miles.'

Asteroid number 3 was discovered in 1804, and named Juno: 'The planet Juno is of a reddish colour, and is free from the nebulosity which surrounds Pallas.' Finally Vesta, discovered by Olbers in 1807: 'Its light is more intense, pure and white than any of the other three. It is not surrounded with any nebulosity, and has no visible disk.' Schröter estimated the apparent diameter as 0.488 seconds of arc, 'one half of what he found for the fourth satellite of Saturn' (Dione).

William Herschel had been in no doubt as to the nebulosities surrounding Ceres and Pallas. In 1802 he recorded that with a power of 550 'Ceres is surrounded with a strong haziness', while Pallas showed up like a 'much compressed, extremely small, but ill-defined planetary nebula'. But later observations failed to show any trace of atmospheres, and the idea that both Ceres and Pallas may have collided with comets, and picked up gaseous surrounds during the encounter, was soon abandoned. Even Ceres, by far the most massive member of the swarm, has much too low an escape velocity to retain an atmosphere of any kind.

*Brewster actually says '1807', but this is an obvious error for 1802.

5   BASIL RINGROSE: ASTRONOMER AND PIRATE

It may not be fair to describe Basil Ringrose as an amateur astronomer, though he did record some observations. He was most certainly a professional pirate, which surely makes him unique!

He was first heard of in 1680 as a member of a pirate fleet which assembled in the West Indies and attacked the Spanish base of Porto Bello in Panama. He attached himself to a captured Spanish ship, *La Santissima Trinidad,* under the command of Captain Sharp, and acted as navigator as they burned and plundered their way down the coasts of Peru and Chile, arriving back in the Caribbean in February 1682. During this voyage Ringrose observed the annular eclipse of September 22 1680, and from it worked out the longitude of the ship. He described the Magellanic Clouds, and wrote: 'Last night we saw the Magellan Clouds, which are so famous among the mariners of these southern seas. The least of these clouds was about the bigness of a man's hat.'

Comets, too, came under his eye. One of these was the Great Comet of 1680, observed by Newton and Halley under rather more peaceful conditions. Ringrose described it: 'November 19th, 1680. This morning, about an hour before day, we observed a comet to appear a degree north from the bright star in Libra. The body thereof seemed dull, and its tail extended itself 18 or 20 degrees in length, being of a pale colour, and pointing directly NNW.' In fact the 'bright star in Libra' was actually Spica, which is in the adjacent constellation of Virgo, but at least Ringrose was not far wrong, and he followed the comet for some nights as it tracked past Spica. His descriptions agree well with those given by Newton and Halley.

In February 1682 Ringrose changed ships, and in March he arrived at the English port of Dartmouth. Not unnaturally, the Spanish authorities took action, and both Ringrose and Captain Sharp were arrested and brought to trial for piracy. One would have thought that the evidence against them was clear enough, but for some reason or other the court freed them, and left Ringrose time to write his journal. This was his last astronomical foray. In February 1686 he went off on another piratical voyage, but this time there was no happy ending: off the coast of Mexico the ship was attacked by the Spaniards, and Ringrose's career came to an untimely end.

Well, at least he had the satisfaction of seeing both the Magellanic Clouds and a brilliant comet!

6 THE TART-LIKE MOON

It is often supposed that Galileo was the first man to look at the Moon through a telescope. This is not so. Thomas Harriot, close friend of the luckless Sir Walter Raleigh and the equally luckless Earl of Northumberland, was making telescopic lunar observations in 1609, several months before Galileo, and his map of the Moon is frankly better than anything achieved by Galileo. One of his ardent disciples was a Welshman, Sir William Lower, who corresponded extensively with him and was probably the second Englishman to observe the Moon. Apparently he received a 'Cylinder' (telescope) from Harriot, and on February 6 1610 wrote to him:

'I have received the perspective Cylinder that you promised me and am sorrie that my man gave you not more warning. . . . According as you wished I have observed the Moone in all his changes. In the new I discover manifestlie the earthshine, a little before the Dichotomie, that spot which represents unto me the Man in the Moone (but without a head) is first to be seene, a little after neare the brimme of the gibbous parts towards the upper corner appeare luminous parts like starres much brighter than the rest and the whole brimme along, lookes like unto the Description of Coasts in the dutch bookes of voyages. In the full she appears like a tarte that my Cooke made me the last Weeke, here a vaine of bright stuffe, and there of darke, and so confusedlie all over. I must confesse I can see none of this without my cylinder.' And in a letter dated June 21 1610: 'In the moone I had formerlie observed a strange spottedness all ouer, but had no conceite that

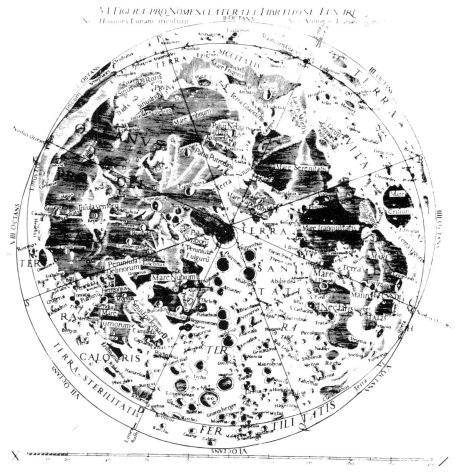

Riccioli's map of the Moon, 1651 — not very like Lower's description!

anie parte thereof might be shadowes; since I haue obserued three degrees in the darke partes, of which the lighter sorte hath some resemblance to shadinesse but that they grow shorter or longer I cannot yet pcaeue.'

This is hardly scientific, and Lower himself is rather a shadowy figure. He was born at St. Winnow in Cornwall in 1570, and entered Exeter College, Oxford, matriculating in 1586. He was admitted to the Middle Temple on August 6 1589, but in February 1591 he was expelled for 'abusing the Master of the Bench following some rowdiness on the previous Candlemas night'. This does not seem to have affected his career. In 1601 he was elected Member of Parliament for Bodmin, and from 1604 to 1611 sat as Member for Lostwithiel; he had been knighted in 1603. He married Penelope, the only daughter and heiress of Thomas Perrot of Treventy, a small mansion in the parish of St. Clears in Carmarthenshire —about nine miles from Carmarthen itself. After his marriage he settled at Treventy, which his wife had inherited from her father, where he seems to have devoted a considerable part of his time to astronomy. He died on April 12 1615, leaving one son, Thomas, who died in 1661. William Lower, the dramatist, was the astronomer's nephew, and on Thomas' death became Sir William's sole heir.

That, really, is about as much as we know, though his correspondence with Harriot ranged over many subjects. We can hardly regard him as a serious astronomer, but at least he deserves to be remembered for his graphic description of the tart-like Moon!

7 FATHER TACHARD AT THE CAPE

All the great observatories of early times were set up in the northern hemisphere, which was natural enough. Southern-hemisphere astronomy began later. It really began in South Africa, and we must spare a thought for Father Guy Tachard, who called at the Cape of Good Hope on his way to Siam in 1685.

The Dutch had colonised the Cape, and at this time the Governor was Simon van der Stel, whose administration was excellent and who remained in office for 20 years. He welcomed the Tachard party, and Tachard may have been over-anxious to be courteous. In his book *Voyage de Siam des Pères Jesuites* he wrote: 'It was agreed upon that the Fort should render Gun for Gun when our Ship saluted it. This Article was ill explained, or ill understood, by these Gentlemen, for about ten of the Clock my Lord Ambassador having ordered seven Guns to be fired, the Admiral answered with only five Guns, and the Fort fired none at all. Immediately the Ambassador sent ashore again, and it was determined that the Admirals Salute should pass for nothing, and so the Fort fired seven guns, the Admiral seven Guns, and the other Ships five, to salute the Kings Ship, which returned them their Salutes, for which the Fort and Ships gave their Thanks.'

The Small Cloud of Magellan.

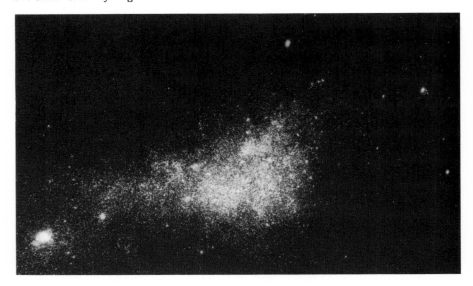

This slight problem having been overcome, Father Tachard and his party went ashore, to be greeted with the utmost cordiality. Tachard wanted to measure the exact longitude of the Cape, and for this purpose he planned to use a method involving the movements of the four bright satellites of Jupiter. He erected a telescope and set to work. Over the next ten days he continued his observations, and also made some notes about double stars—the first telescopic observations of them to be made from South Africa. He detected the double nature of Acrux, the brightest star in the Southern Cross, and also described the two Clouds of Magellan. The Ambassador joined in the observing; and when the time for departure came, the Dutch presented the travellers with 'Presents of Tea and Canary Wine', while Tachard left his hosts a microscope and a burning-glass. All in all it must have been a very pleasant visit; and though Father Tachard never returned to the Cape, he must have carried away a very favourable impression of South Africa and its people, both white and black.

At least he had made a start. Now, three centuries later, the Cape Province is the site of one of the major observations of the world.

8   THE EQUATORIAL SKY

In 1979 I paid a visit to the Philippine Islands, which, as you will recall, lie only slightly north of the Equator. From there I had my first view of the equatorial sky, and I found it quite fascinating.

From England, we miss some of the most interesting objects in the sky. The Southern Cross never rises; neither do Canopus, Achernar, the Clouds of Magellan or the eccentric variable Eta Carinæ. From southern countries such as Australia,

The Eta Carinæ area, with its glorious star clouds.

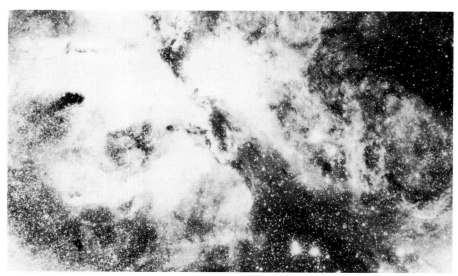

New Zealand and South Africa the situation is better; observers there lose only the northernmost parts of the sky, which are far less brilliant than those near the south celestial pole. The only real losses are the Pole Star, the 'W' of Cassiopeia, and most or all of the Great Bear. Of course, the view is different. From England we see Orion with Betelgeu at the top left and Rigel at the lower right; the three stars of the Hunter's Belt point upward to Aldebaran, downward to Sirius. From Australia, Rigel is higher than Betelgeu, and the Belt points downward to Aldebaran and upward to Sirius.

From the Philippines, however, or anywhere very near the Equator, the two celestial poles lie on opposite horizons, with the celestial equator passing overhead; one can see the entire span of the heavens at one time or another. From the Philippines, Polaris is just above the horizon and the south pole just below, but that is a minor modification.

To an observer exactly on the Equator, it would in theory be possible to see the two pole stars (Polaris and Sigma Octantis) at the same time, though I am not sure that this has ever been done; Sigma Octantis is so faint that it would be very hard to see so low in the sky. The declination of Polaris is N 89° 15′, and of Sigma Octantis S 89° 08′, so that of the two Polaris is slightly closer to its polar point. If anyone has in fact managed to see them simultaneously, I would be interested to know!

Obviously, the planets are on view whenever they are sufficiently far from the Sun, so that in many respects the equatorial observer is very well placed, because there is virtually nothing which is out of view. This being so, why have no major observatories been set up at the Equator? The answer is that everything depends upon local conditions, not only meteorological but also political. Certainly, though, the equatorial sky is of special interest, and it seems curious to see the Southern Cross and the Great Bear at the same time, with Orion lying in an undignified manner on his side.

9   THE GOTTORP GLOBE

Visiting a planetarium is always an interesting experience. Today they are common enough, though the modern type of planetarium date only from 1923, when the first one was set up at the Zeiss Works in Jena, today in East Germany. For three years in the mid-1960s I was Director of the Planetarium at Armagh, in Northern Ireland, so I became very used to giving displays. Of course, there were some awkward moments. I remember that on one occasion I had just started a presentation about 'The Southern Stars' when a circuit fused, and all the southern stars went out. I hardly think that the resulting presentation was one of my best.

There was one interesting 'ancestor' of the present-day planetarium. It was made in Denmark around 1660 by Andreas Busch, a mechanic employed by Duke Frederik III of Holstein, together with the Court mathematician, Olearius. It was known as the Gottorp Globe.

Olearius and Busch realised that to give a real impression of the night sky, the audience must look upward. Accordingly, Busch made a hollow globe 11 ft in diameter, weighing almost 3½ tons, which could be rotated; it was turned by water-power, and spun round once in 24 hours, though the demonstrator could

The Goto projector at Armagh Planetarium in 1967, when I was still Director.

always speed it up if necessary. Inside the globe, hung from its fixed axis, was a round table, with a bench large enough to accommodate ten people. The stars and constellation figures were shown on the inside of the sphere, and the globe was lit from within by two lamps. When the globe was rotated the audience could see the stars drifting across the 'sky'.

The Gottorp Globe proved to be popular. In 1715 it was given to the Czar of Russia, Peter the Great, who in turn gave it to the Academy of Sciences in St.

Petersburg (now Leningrad). Later it was damaged by fire, but was rebuilt and improved. It is still to be seen in Leningrad.

This was a major step forward, and it was soon imitated. Erhard Weigel, at Jena, made a globe which was about the same size as the Gottorp but was made of iron sheeting; the stars were produced by small holes in the sheets, so that when the outside lights were switched on—leaving the interior of the globe in darkness—the audience could see the stars as 'pin-holes' in the dome, the brightness of the star depending upon the size of the hole.

There was, however, one modification which must have been rather alarming. A working Earth was set up inside the globe, with reproductions of the volcanoes Etna and Vesuvius, which gave off steam, flames and what were described as 'pleasant odours'. Hail, wind, thunder and lightning were added for good measure. If all these special effects were turned on at once, we may suppose that the spectators emerged in a somewhat dazed condition—but at least they had had their money's worth!

10 KING GEORGE III AND KEW OBSERVATORY

The Royal Greenwich Observatory was founded by direct order of King Charles II, mainly so that a new star-catalogue could be drawn up for the use of British sea-navigators. Much later, another Royal observatory came into being at the behest of a very different monarch: George III, who, as the history books tell us, reigned officially from 1760 to 1820, though during the last ten years of his life he was hopelessly insane and his son, later George IV, acted as Regent.

An old print of the Old Royal Observatory, Greenwich Park.

Kew Observatory as I photographed it in 1982. Plans are being made to re-instal a small telescope there.

This second observatory was a very modest affair. Though known as the Kew Observatory, it is actually in Richmond. It was set up in the grounds of the Palace of Kew, residence of George III's mother, Augusta, Princess of Wales. By the time that George ascended the throne, the building was derelict, and George decided to have a new Observatory built in the Old Deer Park. It was designed by Sir William Chambers, and was ready by 1769. In that year there was a rare phenomenon—a transit of Venus across the face of the Sun; and the King, whose interest in astronomy was perfectly genuine, was anxious to watch it. A small but adequate telescope was installed, and the transit was well seen.

The first superintendent of Kew Observatory was Dr Stephen Demainbray. He died in 1782, and there were suggestions that he might be succeeded by William Herschel, who had become famous for his discovery of Uranus in the previous year. This did not happen because the King had already promised the post to Demainbray's son. Nevertheless, George did not neglect Herschel. He made him King's Astronomer at a salary of £200 a year, which was enough to enable Herschel to abandon music as a career. The condition was that he should move nearer to the Royal residences. Even before leaving Bath for his new home in Datchet, he visited Windsor, taking a telescope with him, and in a letter written to his sister Caroline on July 3 1782 he recorded that 'My Instrument gave a general satisfaction; the King has very good eyes and enjoys Observations with the Telescopes exceedingly.'

Later, George provided £4,000 to finance the building of Herschel's largest reflecting telescope, with a mirror 49 in in diameter—much larger than any previously made (it remained so until the completion of Lord Rosse's 72-in in 1845).

George IV showed no interest in astronomy; William IV did, and often visited Kew. But finally, in 1841, the Government came to the conclusion that the Observatory could no longer be maintained. It was dismantled, and since then has gone through various transformations, including a period as a seismological station. At the time of writing (1984) it has been acquired by a commercial firm, though it is hoped to retain a small telescope and an exhibition in the old dome, which will still rotate.

It cannot be said that Kew Observatory contributed much to astronomy, but it is worth remembering as being the second observatory to be founded at the wish of a King of England.

11 COMET WINE

It is often said that the most brilliant comets of modern times were those of 1811 and 1843. Sir Thomas Maclear, who saw both, considered that of 1843 to be the more brilliant, but both were visible in broad daylight.

The 1811 comet was discovered by the French astronomer Flaugergues on March 26, and soon became striking. There is a vivid description of it from the pen of Admiral W.H. Smyth, later to become famous as the author of the book *Cycle of Celestial Objects*, but then a serving officer in the fleet off Cape Sicie under the command of Lord Exmouth, the aim being to blockade the port of Toulon. (Incidentally, the Russians later commented that they regarded the comet as a signal indicating that Napoleon's invasion of Russia would be unsuccessful; in this they were right, though one must doubt whether the comet had anything to do with it.) Smyth wrote:

'This splendid object was extremely interesting, not only from its appearance, but from the length of time that it remained visible, nearly 10 months, which was longer than any other on record; and, therefore, none has afforded such certain means of information with respect to its aspect, and its orbit, especially during the time it was circumpolar, and therefore in constant apparition. Both Sir William Herschel and Schröter inferred that it shone by inherent light from the variations in the brightness of the nucleus, and the rapid coruscations of its tail; but that opinion was not generally received. When this comet was in perihelion, its distance from the Sun was about 98,000,000 of miles, and 140,000,000 from the Earth. The gaseous envelope of its head was 30,000 miles thick, and the centre of the nucleus was separated from the interior surface of the surrounding discus, the *head veil* of Schröter, by a space of 36,000 miles; while its central brilliant spot was estimated at 500 miles in diameter. Its tail was composed of two diverging beams of pale light, slightly coloured, which made an angle of 15 or 20 degrees, and sometimes much more; both of these were a little bent outward, and the space between them was comparatively obscure. This tail varied at its greatest extension, from nearly as long as the distance of the Sun from the Earth to 130,000,000 of geographical miles of length; and some observations made it even longer.'

Certainly the comet must have diverted Smyth's attention from the business of battling against Napoleon, but there was another episode which should be remembered. In 1811 the wine vintage in Portugal was particularly good, and the growers attributed this to the comet. 'Comet Wine' was on sale for many years

The bottle of Comet Wine sold at Sotheby's in 1978. Surely this must be the last bottle in existence — I wonder what it would taste like?

afterwards, in the price-lists of port-wine merchants and in auctions. As recently as the 1880s an advertisement for it appeared in the London *Times*.

Rather naturally, I assumed that all the vintage must have been used up long since, but I was wrong. In 1980 I had a letter from Sotheby's, the famous London auctioneers. They had a bottle of Comet Wine for sale; did I know anything about it? Luckily I did, and sent them the information. They were kind enough to photograph the bottle before the auction, and this must, surely, be the last one of all.

I wonder who bought it, and what it tasted like? Or does it still exist? At any rate, it provides a link, albeit a tenuous one, between the space age and the Napoleonic Wars!

12 THE MEANING OF MAGNITUDE

When the Moon is full, its brilliant radiance drowns all but the brightest stars—though it is quite wrong to suppose, as some people do, that full moonlight can rival sunlight. The magnitude of the full moon is about $-12\frac{1}{2}$, while that of the Sun is nearer -27.

'Apparent magnitude' is a measure of the apparent brightness of an object. The scale works rather in the manner of a golfer's handicap, with the more brilliant performers having the lower magnitudes. Thus magnitude 1 is brighter than 2, 2 brighter than 3, and so on.

The four brightest stars in the sky have magnitudes of below zero, so that they have minus values. Pride of place goes to Sirius, at -1.4. Its only rival is Canopus, at -0.7. The other 'minus' stars are Alpha Centauri in the far south of the sky (-0.3) and the northern Arcturus (-0.1). Next in order come three stars just fainter than zero magnitude: Vega and Capella in the northern hemisphere, and Rigel in the southern (though Rigel, in Orion, is not so very far south of the celestial equator, and is visible from every inhabited continent).

Conventionally, the first-magnitude stars are taken as those ranging from Sirius down as far as Regulus in Leo (the Lion), whose magnitude is 1.4. The list of first-magnitude stars includes Procyon in the Little Dog, Achernar in Eridanus, Aldebaran in the Bull, Betelgeux in Orion and so on. The Pole Star is almost exactly of magnitude 2, though admittedly it is very slightly variable.

The faintest stars normally visible with the naked eye on a clear night are of magnitude 6. Binoculars will extend the range down to at least 9, and the world's largest telescopes can record objects down to around magnitude 24.

Note, please, that I am talking about 'apparent magnitude', because a star's apparent magnitude is no firm clue to its real luminosity. The stars are at very different distances. Thus Sirius looks considerably brighter than Canopus, but only because it is nearer. Sirius has 26 times the luminosity of the Sun, while according to a recent estimate Canopus could equal no less than 200,000 Suns; if it were so close to us as Sirius (8.6 light years) it would cast strong shadows. Of the planets, Venus can attain magnitude -4.4, Mars -2.8 and Jupiter -2.6, so that all seem much brighter than any of the stars. Even Saturn, with a maximum magnitude of -0.3, can outshine all the stars apart from Sirius, Canopus and Alpha Centauri.

It is true that 'apparent magnitude' can be rather misleading, but it has been used for many centuries and is not likely to be altered now. Once one becomes used to the system, it is convenient enough.

13   THE WORLD'S MOST UNCOMFORTABLE SPACE-SHIP

Early space-ships were painfully cramped. Compare modern vehicles, such as Spacelab, with the original, one-crew midgets which took up pioneers such as Yuri Gagarin and John Glenn, and the difference is striking. But at least these space-craft worked. The most bizarre and potentially most uncomfortable of all such vehicles was devised by a German inventor, Hermann Ganswindt, as long ago as 1891.

Ganswindt was, quite frankly, an oddity. He was not without ability, and he had some good ideas, but all in all he seems to have been the perfect example of an eccentric professor who produced weird and wonderful contraptions which were at least a source of entertainment. There was, for instance, his helicopter. The first aim of a helicopter is to fly. Ganswindt's did not. He planned to power it by falling weights (of course, internal combustion engines were in their infancy at that period), and its range was limited to a few feet. Another bright idea was his 'tread-motor', a sort of bicycle which was propelled by the rider shifting his weight abruptly from one 'tread' to the other; when he tried it out, it caused considerable consternation!

But it was his space-ship which really caught the imagination. The fuel was nothing more nor less than dynamite. The dynamite was packed into steel cartridges which were fed into a combustion chamber, one after the other, and detonated. Half of the cartridge would leap upward and hit the top of the chamber, thereby giving it a kick which would make it rise from the ground. The spent part of the cartridge would simply fall down, while the next was detonated in its turn and the space-craft would be given a further upward impetus. The cartridges were stored in two columns, one to either side of the combustion chamber, and the crew were stored in a cylindrical cabin underneath the main unit. This cabin had a central hole, through which the spent cartridges could drop harmlessly away. According to Ganswindt, this procedure would involve an upward flight which would be somewhat jerky (!) but would suffice to send the vehicle beyond the top of the atmosphere en route for the Moon. . . . Needless to say, the plan remained on the drawing-board, and no vehicle of this design was actually built. Had it been constructed and tested, one hates to think what the result would have been.

Ganswindt was disappointed, and he had the usual eccentric's view that he was being persecuted by orthodox scientists. Much of his later life was spent in lawsuits which he took out against those who, he considered, had stolen his ideas. He was still busy in these activities when he died on October 25 1934.

However, he did have one real achievement. It was he who invented the free-wheeling mechanism for bicycles. So next time you put on your trouser-clips and prepare for a ride, remember Hermann Ganswindt!

14   METEOR CRATER

'The most interesting place on Earth.' This is how Svante Arrhenius, the famous Swedish scientist, described the great crater in Arizona, not very far from the town of Flagstaff where Percival Lowell set up his observatory to study Mars. Certainly Arrhenius' remark had some justification; there is nothing in the world quite like Meteor Crater.

From the outer plain the view is not impressive, which makes the crater even more spectacular when you climb the rim and look down into it. Measured from rim to rim its diameter is 4,150 ft, and its depth is 570 ft. Visitors can walk along the rim for some distance, and follow the 'trail' which leads down to the floor, though special permission has to be obtained; it is more of a climb than a scramble. The descent takes less than an hour, but making one's way back takes much longer, particularly under the broiling heat of the Arizona sun.

The first white men to find the crater, in 1871, regarded it as volcanic, which was natural enough since there are volcanoes in the area—notably Sunset Crater, which erupted as recently as 1250. Twenty years after the discovery G.K. Gilbert, one of America's leading geologists, visited the site and classified the crater as 'a steam explosion of volcanic origin'. Incidentally, it was Gilbert who championed

Debris of the Siberian impact of 1908. The objects which look like matchsticks are actually pine trees.

the theory that the large craters of the Moon are due to impacts. He was wrong about the Arizona Crater; he was also, I believe, wrong about the Moon, though that whole question is still open.

At any rate, there is no doubt about the origin of the Arizona Crater. Its age is about 22,000 years, so that it was formed long before men had reached the New World. Probably the original meteorite was around 90 ft in diameter; it struck the ground at a velocity of some 30,000 mph, with a force equalling half a million tons of TNT. All life around the impact point would have been wiped out. On impact, 90 per cent of the meteorite was vapourised; the remaining 10 per cent lies buried under the south-east rim of the crater.

In 1903 the crater was explored by Daniel Moreau Barringer, a Philadelphia mining engineer. Barringer disagreed with Gilbert's theory; he assumed that a meteoritic impact had formed the crater, in which case the object was presumably buried underneath. Iron is valuable, and meteorites are iron-rich. Barringer acquired the land by purchase, formed a company to exploit it and set to work. Mining was started in 1906 in the centre of the floor; Barringer believed that the missile had come 'straight down', so that the middle of the crater was the obvious place to look. He was wrong, and in any case he soon ran into trouble. After a year, he had found nothing. He tried again below the south-east rim, but again he was unsuccessful, and finally, at a depth of 2,000 ft, the drill jammed. Nothing more could be done, and the operations were abandoned in 1921. Spasmodic efforts were made until Barringer died in 1929, but by then it had become clear that the meteorite was out of reach.

Since then, the only attempt at commercial exploitation has been in connection with silica mining, since the silica near the rim is as pure as any in the world. Inevitably there was damage to the crater, and the miners became an unmitigated nuisance. In 1967 the site was accepted as a National Natural Landmark, and all mining operations were banned. This was certainly the right decision.

If you are in Arizona, I strongly recommend that you make time to visit Meteor Crater. I think you will agree that the description given by Svante Arrhenius is anything but an overstatement.

15 THE RINGS OF SATURN: THEN AND NOW

There can be little doubt that Saturn is the most beautiful object in the entire sky. Its system of rings is unrivalled; though both Jupiter and Uranus are ringed, Saturn is in a class of its own.

The rings were first seen by Galileo, but his primitive telescope was not good enough to show them in their true form, and he tended to regard Saturn as a triple planet. It was only in 1656 that the problem was solved by the Dutch astronomer Christiaan Huygens. To establish his priority, he published an anagram, as follows:

aaaaaaa ccccc d eeeee g h iiiiiii llll mm nnnnnnnnn oooo pp q rr s ttttt uuuuu.

Rearranged, these letters make up the Latin sentence: 'Annulo cingitur, tenui,

Right *'Spokes' in Saturn's rings photographed from 32 million miles by Voyager 1 on October 4 and 5 1980.*

Below *Saturn's rings as they appeared in 1921, 1933, 1937 and 1941. The cycle is repeated regularly. During the mid-1980s, the rings are wide open.*

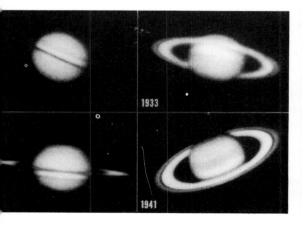

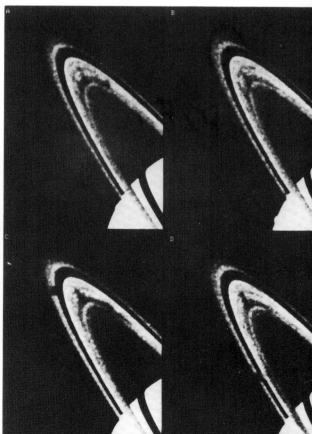

plano, nusquam cohaerente, ad eclipticam inclinato'—which may be translated as '[The planet] is surrounded by a slender flat ring inclined to the ecliptic, but which nowhere touches [the globe].'

Oddly enough, it took some time for this explanation to be accepted, and there were various peculiar theories. Thus Hevelius, a famous astronomer of Danzig (now Gdańsk), believed Saturn to be elliptical, with two appendages attached to its surface. The French mathematician de Roberval suggested that Saturn was surrounded by a hot zone giving off vapours, reflecting sunlight off the edges but appearing opaque when seen in depth. Honoré Fabri, a French Jesuit, claimed that the aspect of Saturn could be explained by the movements of four satellites, two dark and two bright. There was also a theory by Sir Christopher Wren, who, remember, was a professional astronomer before he turned to architecture. According to Wren, Saturn had an elliptical corona, meeting the globe in two places and rotating with Saturn once in each sidereal period. Yet before Wren published his ideas, he read of Huygens' discovery, which by then had been translated. To his eternal credit, Wren promptly accepted it. The Sicilian priest Giovanni Hodierna was not so generous; he believed that Saturn was simply a globe with dark patches on its surface.

Next came a pamphlet by the Italian telescope-maker Eustachio de Divini (actually, it may have been written by Fabri) which went so far as to accuse Huygens of fraud! Huygens replied in humorous vein: 'I had believed that there would be subtle objections unforeseen by me and drawn from profundities of astronomical science with that politeness and modesty befitting a man dedicated to liberal studies. But I was deceived—they attack my observations without solid arguments, accusing me openly of inventing them contrary to truth.'

Telescope improvements soon showed that Huygens was right. Even Fabri withdrew his objections in 1665. All the same, it was a curious episode.

16 POSSIBLE IMPOSSIBILITIES

In Lewis Carroll's immortal *Through the Looking-Glass*, we learn that the White Queen made a daily habit of believing at least six impossible things before breakfast. Scientists, predictably, are much less credulous. All the same, there are times when they go too far in the opposite direction, and a comment by the well-known astronomer and author J. Ellard Gore, made in 1907, provides an excellent example.

In 1862 Alvan Clark discovered the faint Companion of Sirius, which was tacitly assumed to be large, cool and red; it has only $1/10,000$ the luminosity of Sirius itself. In his book *Astronomical Essays*, Gore wrote as follows:

'If its faintness were mainly due to its small size, its surface luminosity being equal to that of our Sun, the Sun's diameter should be the square root of 1,000, or $31\frac{1}{2}$ times the diameter of the faint star, in order to produce the observed difference in light. But on this hypothesis the Sun would have a volume 31,500 times the volume of the star, and as the density of a body is inversely proportional to its volume, we should have the density of the Sirian satellite over 44,000 times that of water. This is, of course, entirely out of the question. . .'. So, opined Gore, the surface of the Companion of Sirius must be much cooler than that of the Sun.

In fact, the density of the Companion is over 50,000 times that of water; it is a White Dwarf, and by no means an extreme case. Gore's 'impossibility' was not impossible at all.

Perhaps the classic case is that of the French philosopher August Comte. In 1830 he made the definitive statement that one piece of knowledge which would be forever beyond the reach of mankind was the chemistry of the stars. Yet within half a century of this prediction, spectroscopy had told us a great deal about stellar chemistry—which goes to show that when a French philosopher makes a profound statement, he is almost certain to be wrong. Incidentally, it was also regarded as absurd to suggest that the Sun could be losing mass at the rate of 4 million tons per second; but it is true enough, though fortunately there is no immediate cause for alarm on this score. The Sun is not likely to change much for at least 5,000 million years in the future.

Finally, let us remember Dr Dionysius Lardner. In an address to the British Association in 1840, he said, gravely: 'Men might as well try to reach the Moon as to cross the stormy Atlantic by means of steam-power.' The North Atlantic voyage was not long delayed, though admittedly the first trip to the Moon took a little longer!

17   THE RIDDLE OF HERCULINA

The asteroids, or minor planets, are small worlds, most of which keep strictly to that part of the Solar System between the orbits of Mars and Jupiter. Only one (Ceres) is as much as 600 miles in diameter, and only one (Vesta) is ever visible with the naked eye. The 532nd asteroid to be discovered was named Herculina. It is faint, and until June 1978 it was thought to be a very ordinary member of the swarm.

Measuring asteroid diameters is not easy, because their apparent diameters are so small; in most telescopes they look like nothing more than starlike points. One new method has been developed by Gordon Taylor, of the Royal Greenwich Observatory. If an asteroid passes in front of a star, it will hide or occult it; and the length of time taken for the asteroid to pass right across the star gives a clue as to the diameter of the asteroid itself. The method works well, the main problem being that one has to wait for a suitable occultation—and Nature has an annoying habit of being perverse.

Taylor found that on June 7 1978 Herculina would occult a dim star, known by its catalogue number of SAO 120774. Observations were made, and we now know that the diameter of Herculina is 135 miles, with an uncertainty of less than two miles either way. But something else was noted. There were two occultations instead of one. This could only mean that there was a second body associated with Herculina.

Herculina may therefore have a satellite, around 30 miles in diameter, which keeps company with it. However, if the satellite has a diameter about one-quarter that of Herculina itself, it might be better to regard the pair as a double asteroid. This was a development which had certainly not been expected. Later in the same year (December 11 1978) a similar phenomenon was observed when a star was occulted by another asteroid, number 18 (Melpomene), which is smaller than Herculina, with a diameter of about 93 miles.

Unforeseen though it was, the existence of asteroid pairs is not really very remarkable. It is now generally believed that the asteroids were formed from material which never collected together to form a large planet because of the disruptive effects of Jupiter, so that pairs could easily have been produced.

What sort of a world is Herculina? Barren; airless; lifeless; probably crater-scarred. It is bleak and inhospitable by any standards, but it is none the less fascinating for that.

18   WHERE JULES VERNE WENT WRONG

Probably the first great space-travel story of near-modern times was *From the Earth to the Moon*, by Jules Verne, which was published in 1865 (its sequel, *Round the Moon*, came out a few years later). In it, Verne's three heroes—Barbicane, Captain Nicholl and Michel Ardan—were fired to the Moon from the mouth of a huge cannon. They went right round the Moon, but their orbit was perturbed by an encounter with an asteroid, so that they eventually fell back to Earth.

Verne was not himself a scientist, but he had plenty of scientific friends, and he believed in making his facts as correct as possible. His speed of departure (7 miles per second) was right; this is the Earth's escape velocity. His fictional telescope, on Long's Peak, was credible enough, though it was more like Lord Rosse's 72-in than the Hale telescope at Palomar. His launching site was not far from Cape Canaveral, and his adventurers plumped down in the ocean much as the Apollo astronauts actually did little more than a century later.

However, Verne did make some mistakes, even though they were not his fault. In particular, the shock of being fired off at escape velocity would be quite a jerk, to put it mildly. In fact the occupants of the projectile would be turned into jelly and killed instantly; moreover, the friction against the thick lower atmosphere would destroy the projectile before it had time to leave the cannon. We are also now certain that there are no asteroids wandering around in the region between the Earth and the Moon, so that using a space-gun would mean a one-way journey only. But the most important of Verne's mistakes concerned weightlessness. In his novel, the travellers became weightless only when they reached the 'zero point' where the Moon's gravity exactly balanced out that of the Earth.

Actually, the travellers would have been weightless as soon as they began their journey—as is the case with the astronauts of today. Also, there is no 'zero point'. Obviously there comes a time when the pull of the Moon is more evident than that of the Earth, but the pull of the Sun is always more powerful than either, and the so-called neutral point is a myth.

Despite all this, Verne's novels were fascinating, and even today, when the Moon has been reached, they have not lost their charm. If you have not read them, I strongly recommend you to do so.

Facing page *In Jules Verne's projectile, en route for the Moon (from an old woodcut in the original French edition).*

19   VENUS IN ACTION

Have you ever heard an eye-witness recording of the Krakatoa eruption of 1883? It must have been terrifying—the old island was destroyed, and the death-roll was colossal. Yet according to new evidence, something much more catastrophic happened on Venus during 1978. This was the time when the US orbiting probe Pioneer began to circle Venus, sending back data and mapping the surface by radar.

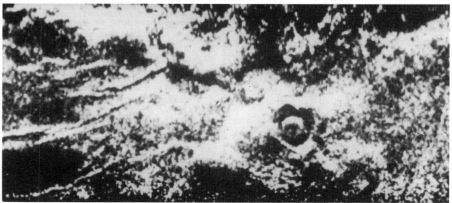

Top *The rotation of Venus as photographed by Pioneer in 1978. The arrows indicate the same regions, confirming the four-day retrograde rotation period of the upper clouds.*

Above *Lava flows on Venus, shown by the Russian orbiter Venera 15 in 1983-4. There is also one prominent crater.*

We were already fairly sure that Venus was a world with active volcanoes. There are two main regions: Beta Regio, made up of two huge shield volcanoes, Rhea Mons and Theia Mons; and Atla Regio, some way away. (The even larger and taller volcano, Maxwell, seems to be extinct, or at least dormant.) From studies over 1,000 Pioneer orbits, Drs L. Esposito and F. Scarf, in America, have found evidence for clusters of lightning bolts over Beta and Atla, similar to those discharges generated in terrestrial volcanic plumes. What seems to happen is that Beta and Atla sit over powerful upflowing convective plumes deep in Venus' interior magma. These plumes or 'hot spots' burn their way through the thick, rigid crust of the planet. Though Venus and the Earth have probably similar amounts of internal heat, Earth has mobile 'plates', and the heat can escape in many places more or less continuously. Venus has not, so that Beta Regio and Atla are almost the only sources of escape.

The 1978 eruption seems to have forced huge amounts of sulphur dioxide and small haze particles into the atmosphere. The sulphur dioxide in the gas quickly formed into small aerosol particles of sulphuric acid similar to the notorious acid rain on Earth, but far more concentrated. It has been estimated that the sulphur may have been blown up as high as 40 miles above the ground, with ten times the force of any Earth volcano over the past century—even Krakatoa. And after the eruption, the sulphur dioxide in Venus' atmosphere increased from a mere two parts per 1,000 million to as much as 100 parts per 1,000 million. Since then the level has dropped—but for how long? We have less certain evidence of a similar outbreak in the 1950s. . . .

At any rate, both the Russian orbiters, Veneras 15 and 16, and the ground-based radar from Arecibo in Puerto Rico indicate the existence of volcanic cones in Venus, and Beta Regio seems to be associated with bright rays which are presumably very young lava-flows. The more we learn about Venus, the less welcoming it seems to become. Quite apart from the tremendous atmospheric pressure and the corrosive sulphuric acid which is always present in the clouds, to say nothing of a surface temperature over 900 degrees Fahrenheit, we now have to reckon also with unpredictable but unbelievably violent eruptions.

Less than 30 years ago it was still thought that Venus might be a reasonably temperate and calm world, with wide oceans and perhaps even primitive life. We were wrong. Venus is a planet to be studied from a respectful distance!

20   TRANSFERRED STARS

In 1603 the German astronomer Johann Bayer published a star catalogue. It was quite a good one, but is memorable chiefly because Bayer introduced the system of allotting Greek letters to the stars in their various constellations. The first letter of the Greek alphabet is Alpha; therefore, the brightest star in (say) Lupus, the Wolf, should be Alpha Lupi, the brightest star in Lyra (the Lyre) Alpha Lyræ, and so on through to Omega, the last of the Greek letters. The system was a convenient one, and is still used. Only in the cases of the brightest stars are the individual names commonly accepted; thus Alpha Lyræ is better known as Vega.

In some cases the letters are out of order. In Orion, Beta (Rigel) is brighter than Alpha (Betelgeux), and there are many other cases. There are also instances of stars being transferred from one constellation to another. The system accepted today was laid down by the International Astronomical Union in 1932, and will certainly not be altered, but it is interesting to look back and see which stars have been given free transfers.

The first, and most illogical, case is that of Alpheratz, which is of the second magnitude and is one of the four stars making up the Square of Pegasus. Bayer naturally included it in Pegasus, as Delta Pegasi, but the IAU decree shifted it into the adjacent constellation of Andromeda, and it became Alpha Andromedæ (in fact Bayer had given both designations).

The other notable case is that of Al Nath, also of the second magnitude. It was originally included in Auriga, the Charioteer, as Gamma Aurigæ, and this was reasonable enough because Auriga is a well-marked constellation (its leader is, of course, the brilliant Capella). However, Al Nath is now in Taurus, as Beta Tauri. Taurus itself is one of the Zodiacal constellations, and includes not only the bright orange Aldebaran but also the two most famous of all open clusters, the Pleiades and the Hyades (to say nothing of the Crab Nebula, though the Crab is far too

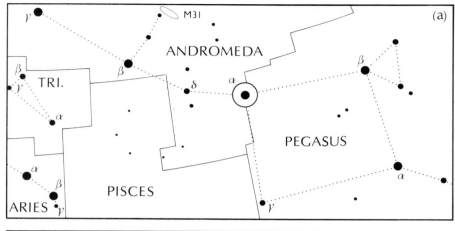

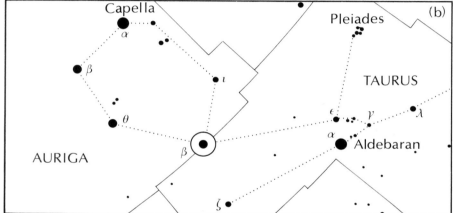

Common stars (circled): (a) Alpha Andromedæ; (b) Beta Tauri; and (c) Sigma Libræ.

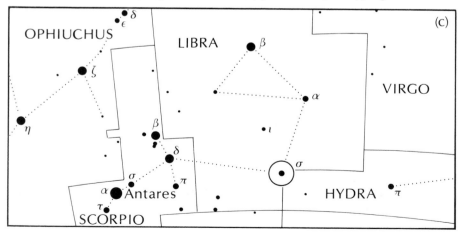

faint to be seen with the naked eye). Taurus is frankly rather shapeless, but the transfer of Al Nath is much less irrational than that of Alpheratz.

Libra, the Balance, is also in the Zodiac, but is faint and formless. Zubenelhakrabi, once known as Gamma Libræ, is now in the Scorpion as Sigma Scorpii (in fact, Libra itself was once Chelæ Scorpionis, the Scorpion's Claws). And Canis Major, the Great Dog, has lost the star lettered 3, which has become Delta Columbæ in the neighbouring constellation of the Dove. Finally, the three main stars of the little constellation of the Shield—Alpha, Delta and Beta Scuti—were formerly 1, 2 and 3 Aquilæ.

It cannot be said that the overall pattern of the constellations is convenient, and if we had followed, say, the Chinese or Egyptian systems everything would be different. However, we have become so accustomed to them that there is no reason for any new changes.

21   HOW THE LUNAR CRATERS WEREN'T FORMED

Even today, after men have reached the Moon and lunar rocks have been analysed in our laboratories, astronomers are still arguing about how the lunar craters were formed. Some believe them to be volcanic (using the term in a broad sense), while others attribute them to meteoritic impacts. No doubt both processes have operated. Meanwhile, there have been some past theories which are rather less rational, but which deserve to be put on record.

One of the strangest, to my mind, was proposed by D.P. Beard in 1920. Beard supposed that the lunar seas had once been water-filled, and that the craters were simply coral atolls which had been built up. On rather the same lines was the Ice Theory, which originated with a tea-planter named Peal but was supported later by several serious astronomers—notably Philipp Fauth, of Germany, author of a large though rather inaccurate lunar map. This time the entire lunar surface was ice-covered, so that the mountains supported huge glaciers and the craters had ice walls. The fact that at the Moon's equator the temperature can rise to over 200 degrees Fahrenheit did not seem to be taken into account—nor was the fact that ice walls would gradually flatten out under their own weight. Surprisingly, Fauth continued to champion this idea up to the time of his death during the last war, effectively ruining his scientific reputation in the process.

I have always been intrigued by a Spanish engineer, Sixto Ocampo, whose thoughts ran along completely different lines. To him, the Moon used to be inhabited. There was a war between two opposing races, and the craters were the scars left by atomic bombs. The fact that some craters have central peaks while others do not proves, of course, that the two sides used different kinds of bombs. The last bombs to be exploded on the Moon 'fired' the lunar seas, which fell to Earth *en bloc* and caused the biblical Flood. Ocampo sent this thesis to the Barcelona Academy of Arts and Sciences, but for some strange reason they refused to publish it. Mortally offended, Señor Ocampo went to America and published it himself, prefacing it with a note in which he claimed that an unscrupulous British astronomer had stolen his theory and was planning to publish it as his own, thereby depriving Spain of the honour and glory. His life's work was done, and he died almost immediately.

Left *Lunar craters. Ptolemaeus is the large, smooth-floored formation at the bottom; Alphonsus immediately above it.*

Below *The Sinus Iridum. The oval, dark-floored crater to the left is Plato, 60 miles in diameter. Between Plato and Sinus Iridum, note the curious Straight Range.*

Right *Tycho, the 54-mile lunar ray crater, seen from orbit.*

Below right *Wide-angle photo taken by Lunar Orbiter 3 showing the prominent crater Kepler and, to its right, the smaller, almost perfectly formed Kepler A.*

It was, of course, difficult to surpass Señor Ocampo, but the Austrian engineer Weisberger managed it by the simple expedient of denying that there were any lunar mountains or craters at all. He attributed the appearance of the Moon's surface to storms and cyclones in a dense, low-lying atmosphere. I met him once, but found it rather difficult to argue with him, as any criticism was taken as a personal insult. However, as he spoke no English and I speak no German, we had to manage in fractured French, and our conversation was limited—which, under the circumstances, was perhaps just as well!

22   UNEXPECTED DISCOVERIES

Delving back in astronomical history, we find various cases of 'serendipity'—finding something which was not expected and which was not being sought. Perhaps the most remarkable case concerns no less a person than Galileo, the first great telescopic observer, who built his tiny 'optick tube' and used it, in the winter of 1609-10, to make a series of discoveries which overturned many long-cherished theories of the universe.

Galileo made an early observation of the four bright satellites of Jupiter: Io, Europa, Ganymede and Callisto. He may not have been the first to see them, but he was undoubtedly the first to study them in detail, which is why they are usually called the Galileans. But there was another fascinating object which he saw twice—once in December 1612 and again in January 1613—though he naturally mistook it for a star. Modern calculations show that this was in fact the planet Neptune, in the same telescopic field with Jupiter. In fact, around that time Neptune was occulted by Jupiter (something which will not happen again until October 27 2088). So Galileo recorded Neptune 233 years before it was positively identified by Johann Galle and Heinrich D'Arrest, in 1846.

Going back further, to the year 364 BC, we come to some observations made by a Chinese star-gazer named Gan De. From the surviving records he not only looked at Jupiter (with the naked eye, of course), but also recorded a 'small star' very close to it. It may well have been Ganymede, the largest and brightest of the Galilean satellites, which would be an easy naked-eye object if it were not drowned by the brilliance of Jupiter. Gan De may have seen two of the Galileans very close together; we cannot be sure, but at least he seems to have the honour of priority, even though he can have had absolutely no idea that he was looking at a satellite rather than a star.

A modern instance is provided by Dr Lubos Kohoutek, a Czech astronomer working at the Hamburg Observatory. There are some comets which have 'gone missing' and have apparently disintegrated. One of these is Biela's periodical comet, which was last seen in 1852. Kohoutek was looking for it in a possible position indicated by calculation when he found, not Biela's Comet, but a new one—Kohoutek's Comet, which never became as brilliant as expected, but which was visible with the naked eye in 1973, and was studied by the crew of the Skylab space-station when it approached the Sun.

Finally, I must record that I have been concerned in one instance of serendipity myself. In 1966 Saturn's rings were edge-on to the Earth, and this was a very

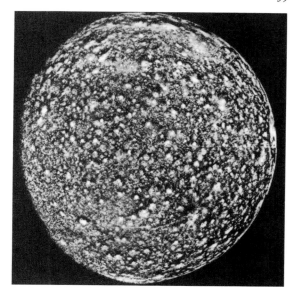

Callisto, Jupiter's outer large satellite. It was discovered in 1610; this photograph was taken from the Voyager 2 probe in 1977, showing Callisto to be covered with craters.

favourable time for studying the faint inner satellites. I was doing so, using a 10-in refractor, when I recorded a dim speck which seems to have been the satellite we now call Janus. I can claim no credit whatsoever, as I did not recognize it as being new, and it was not 'discovered' before the Voyager missions of more than a decade later. But it does show that one must always be on the alert!

23 OUT OF THE SOLAR SYSTEM

I think most people know that by now, every planet in the Solar System as far out as Saturn has been studied by space-craft. Each has provided its quota of surprises. Mercury, with its craters and basins; Venus, with its volcanoes and its deadly acid rain; Mars, with its volcanoes and dry watercourses; Jupiter, with its amazing system of satellites—including Io, the most volcanically active world in the Solar System; and Saturn, with its complex ring-system. Uranus is next on the list, due to be bypassed by Voyager 2 in 1986; and finally Neptune in 1989, leaving only Pluto unexplored. But what will happen to the various probes when they have completed their main tasks?

Let us concentrate on the two Pioneers, 10 and 11. Pioneer 10 was the first probe to encounter Jupiter, in December 1972, after which it began a never-ending journey out of the Solar System. On June 13 1983 it crossed Neptune's orbit, though Neptune was nowhere near it at the time. This means that Pioneer 10 is now the most remote man-made object ever tracked. It will never return; it is travelling too fast, and though we may hope to maintain radio contact with it until well into the 1990s we are bound to lose it eventually.

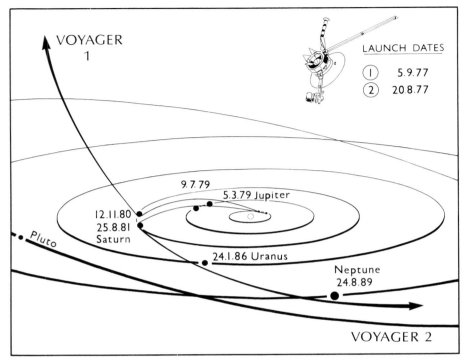

Orbits of the Voyager space-craft.

Pioneer 11 encountered Jupiter in December 1973, and was then swung back across the Solar System to an encounter with Saturn in 1979. It too will never return—and note that Pioneers 10 and 11 are leaving the Solar System in opposite directions. With luck, we may be able to keep in touch with them until they have passed over the boundary of what we call the heliosphere.

The heliosphere is the region of space in which the Sun's influence is dominant. Now, the Sun sends out a constant stream of particles in all directions, making up the solar wind. Of course, the solar wind cannot extend indefinitely. There must be a point where the wind ceases to be detectable, and this marks the edge of the heliosphere.

We have already found that the heliosphere is variable; it expands and contracts according to the state of activity of the Sun. Astronomers are anxious to find out more about it, and the boundary is an important region. The heliosphere is compressed in the direction of the Sun's motion through space, and is stretched out in the opposite direction, so that it is shaped rather like a tear-drop. Pioneer 10 is moving 'downstream' and Pioneer 11 'upstream', so that Pioneer 11 will reach the boundary first.

Neither probe will encounter another star for a very long time. It must happen eventually; for instance, Pioneer 10 may pass within six-and-a-half light-years of Altair in about 227,000 years from now, but so far as we are concerned I am afraid that this is of no more than academic interest!

24 THE MOON ILLUSION

When the Moon is nearly full, anyone can see the famous light and dark markings on its face; some people can make out the so-called Man in the Moon, though I admit that I have never been really successful. The dark patches are the waterless lunar seas, which were once oceans of lava; the bright areas are the uplands. Any telescope, or for that matter binoculars, will show the Moon's mountains and craters. The lunar landscape is of immense beauty as well as interest, and the fact that men have landed there has not in any way lessened the romance of the Moon.

There is something else which has intrigued many people for many years. It is often said that when the full moon is just rising, it looks very large; as it climbs in the sky it shrinks, and when high up it seems comparatively small. Ask the average person whether the full moon looks largest when it is low down, and he is almost certain to answer 'yes'.

Actually, this is not true. The low-down full moon is no larger than the high-up

Full Moon — or nearly so; actually, it is just past full (see left-hand edge).

full moon. If you want to prove this, you can do so easily. Select a small pebble (or something equivalent) and hold it out in front of you until it just covers the rising full moon. Repeat the experiment when the Moon is high up, and you will find that there is no difference.

Yet it is undeniable that the low-down full moon *seems* to be large. This is the celebrated Moon Illusion, and it has been known for centuries.

One man who commented upon it was Ptolemy, last of the great astronomers of the Greek school, who lived around AD 150. His explanation was that the low moon is seen against a background of 'filled space'—trees, houses and so on—while when the Moon is high there is nothing with which to compare it; it is seen against 'empty space'. Hence the illusion.

Some years ago, in one of my BBC television *Sky at Night* programmes, we tried an experiment. With Professor Richard Gregory, we built an apparatus in which we could see the image of the real Moon and compare it with an artificial moon. It was worked by mirrors, and we tested it one evening on the beach at Selsey, much to the amusement of various onlookers. We found that the observer still judged the low Moon to be bigger than the high Moon, though in each case the comparison object was an artificial moon which did not change. We also tried various other tricks. One member of our team covered up one eye for several hours beforehand; the illusion was still present. I even observed while standing on my head, balancing myself precariously against a groyne and creating the general impression that I was insane.

I am never sure whether Ptolemy's explanation is the full answer, but in any case there is no doubt that the effect is an illusion and nothing more. If you do not believe me, experiment for yourself!

25   THE TWILIGHT ZONE OF MERCURY

Of all the 'ancient' planets—that is to say, those which were known by the start of scientific history—Mercury is much the least prominent, and there are many people who have never seen it at all. It is only 36 million miles from the Sun on average, and can be seen with the naked eye only when very low in the west after sunset or very low in the east before sunrise. Moreover, it is only 3,030 miles in diameter.

Mercury takes 88 days to complete one journey round the Sun. Before the space age, it was generally believed that it also took 88 days to spin once on its axis, so that it would keep the same face turned toward the Sun all the time. One part of the surface would be in permanent sunlight, and another part plunged in everlasting night. Yet because the path of Mercury round the Sun is elliptical rather than circular, its orbital speed changes, and this would produce a sort of intermediate belt over which the Sun would rise and set, bobbing up and down over the horizon. In fact, Mercury would have a Twilight Zone.

Behaviour of this sort would not be unexpected. The Moon keeps the same hemisphere turned toward the Earth; tidal friction over the ages has been

Facing page *Two views of Mercury from Mariner 10, showing the heavily cratered surface.*

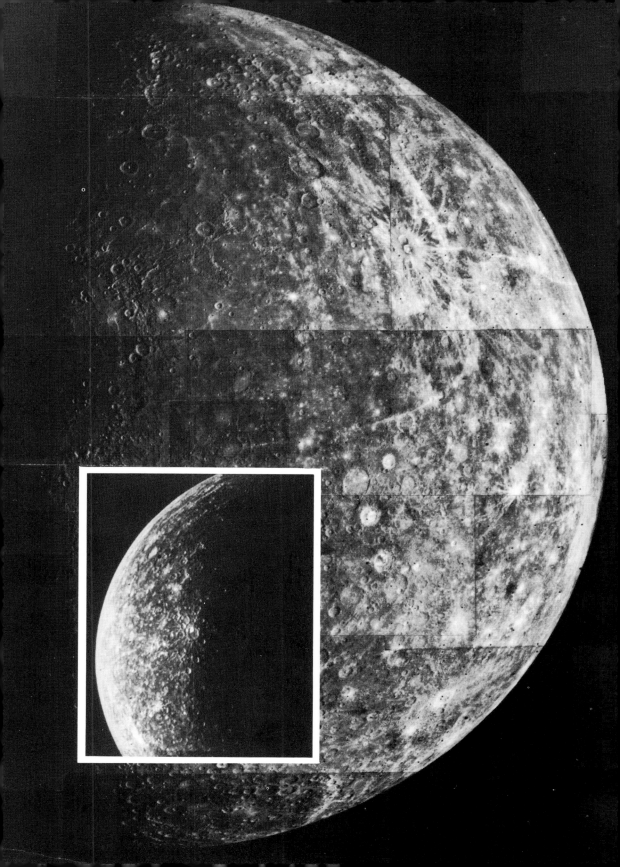

responsible, and there is no mystery about it. Most of the satellites of the other planets have similar 'synchronous' rotations.

Science fiction writers loved the Twilight Zone of Mercury. It represented the only part of the planet which would have a reasonable temperature; the area of permanent sunlight would be intolerably hot, and the darkened area equally intolerably cold. It was said that Mercury was both the hottest and the coldest planet in the Solar System. Could low forms of life flourish in the Twilight Zone, despite the virtual lack of atmosphere? And could manned bases be set up there eventually?

Alas for such speculations! In the 1960s serious doubts arose. The 'night' side seemed to be much warmer than it would be if it never received any sunlight. Finally, in 1974, the problem was solved when the space-craft Mariner 10 flew past the planet and sent back information from close range. The rotation period is not 88 days. It is only 58½ days, which is admittedly slow, and which gives Mercury a weird calendar—but which does away with the Twilight Zone.

Since then, science fiction writers have had to alter their plots. Cratered, airless little Mercury is even less welcoming than we used to think.

26   A BLACK HOLE IN THE SCORPION?

Scorpio or Scorpius, the Scorpion, is one of the brightest constellations in the sky. It is also one of the few groups to bear at least a vague resemblance to the object it is supposed to represent; one can picture a scorpion when looking at the long, curved line of bright stars which make up the main constellation. The leader is the bright red supergiant Antares. However, there is an object in the Scorpion which cannot be seen with the naked eye, but which seems to be of exceptional interest. It is known as V.861 Scorpii, and lies close to the third-magnitude star Zeta Scorpii.

At first sight there is nothing unusual about V.861, but every 7.8 days it 'winks', dimming down for a while before recovering its lost light. There is no mystery about this; V.861 is an eclipsing binary—that is to say, a system made up of two stars moving round their common centre of gravity. When the brighter member is hidden by the dimmer companion, the light which we receive is partly cut off. Many eclipsing binaries are known; Algol, the 'Winking Demon' in Perseus, is the most famous of them.

Recently, however, it has been found that V.861 is also a source of X-rays. Now, these X-rays are of very short wavelength, and are difficult to study from ground level because the Earth's atmosphere blocks them out. We have to study them from space, and a great deal was learned from the instruments carried aboard the artificial satellite 'Copernicus', which was sent up specifically to study X-ray sources. With V.861 the X-rays are also cut off periodically, and it now seems that they come from the dimmer member of the pair, which we cannot actually see.

To make this clearer: there are two stars in the system—one which we can see, and one which we cannot. The visible star sends out light, and when the fainter member passes in front of it it gives a slow 'wink'. The invisible star sends out X-rays; when the bright member hides the faint one, it is the X-rays which are cut off.

We must ask, then, whether the invisible member is a normal star. The mass is thought to be between five and 12 times as great as that of the Sun, and this indicates that the object may be a Black Hole—in other words, a very old, massive, collapsed star which is pulling so powerfully that not even light can escape from it. If so, the X-rays come from material which is intensely heated just before it is sucked into the black hole.

Time will tell whether or not this interpretation is correct, but at any rate there is no doubt that V.861 is a particularly interesting object, and it will be intensively studied in the years to come.

27   FAREWELL TO THE BEER CONTINENT

The first attempts at mapping the surface of Mars were made in the early 19th century, and by around 1860 the main bright and dark areas had been fairly well defined, though it was believed that the dark areas were seas (or, at least, sea-beds filled with vegetation). Not unnaturally, names were given to the various features, and the system of naming Martian markings after astronomers was adopted. Thus one of the reddish areas was named Beer Continent in honour of Wilhelm Beer, one of the first really good lunar and planetary observers (he collaborated with his even more famous contemporary, Johann Mädler). The idea of a Beer Continent is intriguing! Other astronomers were also given lands or seas of their own, but unfortunately different observers used different systems, and everything became so chaotic that in 1877 G.V. Schiaparelli, observing from Milan, decided to scrap the whole system and begin all over again. With modifications, it is Schiaparelli's

Map of Mars by Camille Flammarion, showing one of many variants of pre-1877 nomen-clature.

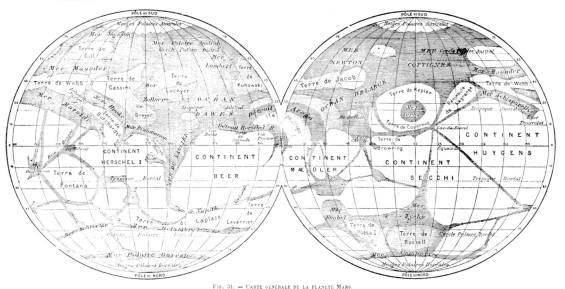

Fig. 31. — Carte générale de la planète Mars.

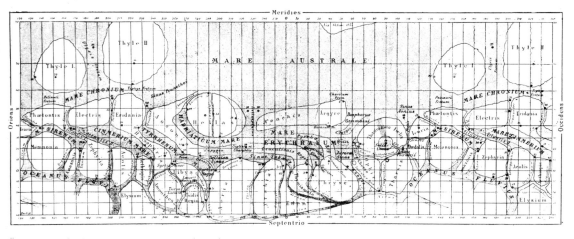

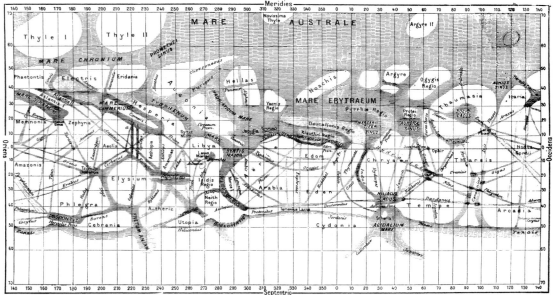

system which we use today. But it is worth looking back at some of the names used before 1877, and which did not entirely die out until the end of the century.

Arago Strait Now Margaritifer Sinus, the Gulf of Pearls—off the coast of South India.

Beer Continent Now Æria, the Greek name for Egypt, 'the far land of mist'.

Bessel Lake Now Phœnicis Lacus. The phœnix was a mythical bird which burned itself to ashes every 500 years, only to rise again as good as new.

Dawes' Forked Bay Now Sinus Meridiani, or Meridian Bay, long used as the zero for Martian longitudes.

De la Rue Ocean Now Mare Erythræum, an old name for the Indian Ocean.

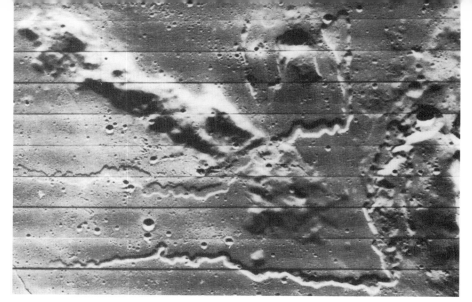

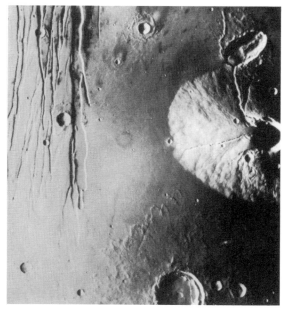

Above left *Schiaparelli's 1877 map of Mars, showing 'canali' and giving his revised nomenclature.*

Left *Schiaparelli's later map of Mars following his observations from 1877 to 1886.*

Above *Old riverbeds on Mars, seen from Viking.*

Right *Ceraunius Tholus, one of the Martian volcanoes, as photographed from Viking. Note the lava-flows and the large summit caldera.*

Flammarion Sea Now Syrtis Minor, after a bay on the Mediterranean coast of North Africa.

Herschel I Continent Now Æolis, named after the floating island which was the home of Æolus, god of the winds. (Herschel I was William Herschel, to distinguish him from his son Herschel II (John). Both certainly merited being included.)

Herschel II Strait Now Sinus Sabæn, the Gulf of Aden. Saba was an incense-rich area in the south of the Arabian peninsula.

Hooke Sea Now Mare Tyrrhenum, the sea bordering Italy.

Hourglass Sea or *Kaiser Sea* Now Syrtis Major, the most prominent of all the

48

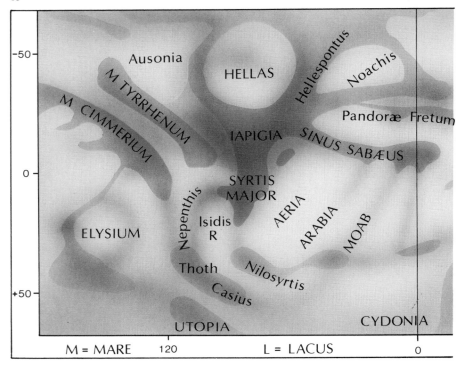

Map of Mars showing old place names.

-50

0

+50

Ausonia
HELLAS
Hellespontus
Noachis

M TYRRHENUM
M CIMMERIUM
IAPIGIA
SINUS SABÆUS
Pandoræ Fretum

SYRTIS MAJOR
Nepenthis
Isidis R
AERIA
ARABIA
MOAB

ELYSIUM

Thoth
Nilosyrtis
Casius

UTOPIA
CYDONIA

M = MARE   120   L = LACUS   0

Above *Map of Mars showing old place names.* **Right** *The Phaethontis area of Mars from Mariner 9, January 1972. At this time Mariner was 1,282 miles from the surface.*

dark areas; a Libyan gulf. The name 'Hourglass' was widely used because the dark area really does have this shape—provided that one uses a little imagination.

Kepler Land Now Thaumasia, after Thaumas, a god of the clouds.

Knobel Sea Now Mare Acidalium. Acidalius means 'referring to Venus'. In the latest nomenclature, following the space probe results, Mare Acidalium has become Acidalia Planitia.

Lockyer Land Now Hellas, or Greece.

Mädler Continent Now Chryse, the 'Golden Plain', the site of the landing of the Viking 1 probe.

Maraldi Sea Now Mare Cimmerium, the island home of an ancient Thracian seafaring people, 'wrapped in mist and cloud'.

Phillips Island Now Deucalionis Regio, after Deucalion, son of the mythological King Prometheus of Thessaly.

Schiaparelli Sea Now Mare Sirenum, the Sea of the Sirens—in mythology, beautiful maidens who sang so sweetly to passing mariners that the ships were lured on to the rocks. (This, by the by, was not a name given by Schiaparelli himself. It was used by the French astronomer Camille Flammarion and others.)

Secchi Continent Tharsis; possibly the ancient Spanish town of Tartessus, destroyed in 500BC . The Tharsis Ridge is the site of the giant Martian volcanoes.

Terby Sea Now Solis Lacus, the Lake of the Sun, often nicknamed 'the Eye'; it

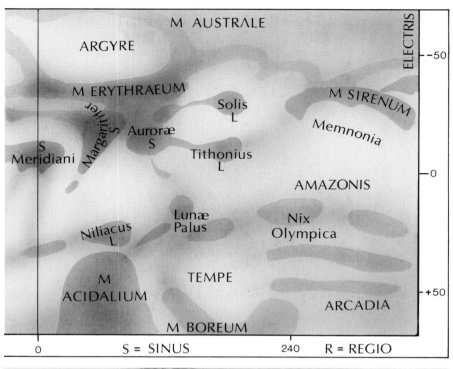

M AUSTRALE

ELECTRIS

ARGYRE

−50

M ERYTHRAEUM

M SIRENUM

Solis
L

Margaritifer

S Auroræ
S

Memnonia

S
Meridiani

Tithonius
L

AMAZONIS

0

Niliacus
L

Lunæ
Palus

Nix
Olympica

M
ACIDALIUM

TEMPE

+50

ARCADIA

M BOREUM

0   S = SINUS   240   R = REGIO

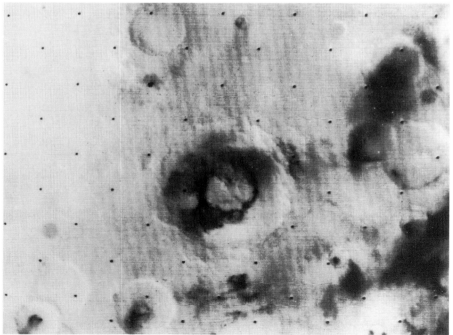

was originally used as the zero for Martian longitudes, and appears more prominently on the old maps than it seems today.

Webb Land Now Phæthontis, after Phæthon, son of the Sun God who was allowed to drive the sun chariot for one day and was struck by a thunderbolt when he showed imminent signs of losing control and setting the world on fire.

Our views of Mars have changed. For instance Lacus Phœnicis, the 'Phœnix Lake', is not a lake but a huge volcano. No doubt the names of today are much more suitable, though some may regret the loss of Beer Continent and the Hourglass Sea.

28   ADMIRAL SMYTH AND HIS DOUBLE STARS

In 1844 a book called *Cycle of Celestial Objects* was published by a retired naval officer, Admiral William Henry Smyth, who had set up a private observatory at Bedford and equipped it with an excellent 5.9-in refractor. The book (or, rather, the second part of it) contained many measurements of double stars. Smyth gave the angular distances between the components, and also the position angles—that is to say, the direction of the fainter component with reference to the brighter. In the following year the Royal Astronomical Society awarded Smyth its Gold Medal, and elected him President for a two-year term. His appointment was popular; his term as President was a distinct success and his double-star measurements were regarded as the best of their time.

Smyth died in 1865. Then, in 1879, an English astronomer, Herbert Sadler, published a paper in the *Monthly Notices* of the Royal Astronomical Society in which he claimed that many of Smyth's results had been merely copied from previous published observations. In fact, the *Cycle* itself was worthless, in which case Smyth had been guilty of the gravest crime which a scientist can commit—faking observations (shades of the Chevalier D'Angos, of whom more anon).

There was an immediate outcry. Smyth had been well liked everywhere; to regard him as a charlatan was unthinkable—and yet Sadler was right; many of the *Cycle* positions were wrong.

The leading double-star observer of that period was S.W. Burnham of the United States. He was consulted and agreed to re-examine all Smyth's pairs using the fine 18½-in refractor at Chicago. Before long Burnham solved the problem. Smyth's observations were of two kinds: measurements of wide, well-known pairs, for which his results were fairly good, and measurements of much closer pairs, not previously examined, where Smyth's figures were badly in error.

What Smyth had done was to give mere estimates for the wide pairs—but unfortunately he had presented his work in a way which, quite innocently, gave the impression that the measurements were his own. The values for the fainter and closer pairs depended upon observations made by Smyth himself, and were very inaccurate. Moreover, he had made frequent mistakes in calculation and in the reduction of the measurements. His estimated colours for the double-star pairs were also wrong. Therefore, Smyth was vindicated inasmuch as he was completely honest, but Sadler had been right in claiming that his measurements were of little value.

It is a sad story; but nevertheless, the *Cycle* still makes fascinating reading. It has been reprinted several times; I have a copy of the last edition, which was published in 1881. Enjoy it by all means—but do not make the mistake of relying upon it.

29   HOLMES' COMET: A WANDERER RECOVERED

Comets have short lives on the cosmical scale. Each time a comet returns to perihelion (that is to say, its closest point to the Sun) some of its icy material evaporates and it loses mass. Periodical comets, which reach perihelion every few years, suffer constant wastage, and several comets which were observed during the last century have now disappeared—Biela's, Brorsen's and Westphal's, for example. But it is never safe to assume that a comet is dead because it remains unobserved for many years, and this was brought home forcibly in recent times by the case of Holmes' Comet.

In 1892, E. Holmes discovered a comet which could just be seen with the naked eye, and which had a period of rather over seven years. It came back regularly on schedule, but at the return of 1908 it was much fainter than it had been when Holmes discovered it. Then it 'went missing', and for a long time it could not be found. Most astronomers thought that it, like Biela's Comet and others of the kind, had died.

When working out the orbit of a comet, every effort has to be made to allow for the gravitational pulls of the planets. Since a comet is a flimsy object, of slight mass, it can be perturbed very easily. Brian Marsden, an English astronomer working in America, set to work to see whether he could track down Holmes' Comet. He decided where it ought to be at the return of 1964—and, sure enough, there it was, though instead of being fairly bright it had become extremely faint. It had been recovered after a lapse of over half a century.

This is not the only case of the rediscovery of a long-lost wanderer. Denning's Comet of 1881, which has a period of nine years, was not picked up again until it was recovered in 1978 by the Japanese observer Fujikawa. But there are interesting comets which have not been recovered, despite intensive searches for them. One is Wilson-Harrington, seen for the first and only time in 1949. It was calculated to have a period of 2.3 years, which is shorter than that of Encke's Comet, which, with its period of 3.3 years, has the shortest period known. Encke's Comet has now been seen at more than 50 returns. It was originally found by the French observer Méchain in 1786, seen again by Caroline Herschel in 1795, by Thulis in 1805 and by Pons in 1818. The orbit, was calculated by Johann Encke, who successfully predicted a return for 1822 (which is why the comet is named after him). Encke's Comet can now be followed throughout its orbit, and it certainly will not be lost; but Wilson-Harrington has vanished. Perhaps it has disintegrated; perhaps, like Holmes' Comet, it will be found again one day.

30   THE STRANGE STAR IN THE WOLF

Lupus, the Wolf, is one of the less famous constellations. It is too far south to be seen from most of Europe and it contains no really bright star; neither does it have a distinctive pattern. But it has one distinction: in the year 1006 it was the site of what appears to have been the brightest supernova ever observed.

Supernovæ are of two types, which may have different origins, but in one class of supernova we see the death of a formerly massive star. It has been shining because of nuclear transformations going on inside it; when these fail, the star begins to collapse. There is a tremendous implosion (the opposite of an explosion), followed by a shock-wave which disrupts the star and sends most of its material flying away into space in all directions. At its peak, the outburst gives off at least 15 million times the energy of our Sun. The end product of this kind of supernova is a cloud of expanding gas, sometimes with the remnant of the original star showing up as a tiny, incredibly dense object made of neutrons—that is to say, atomic particles with no electrical charge. Neutron stars are generally known as pulsars, because they spin round very quickly and emit pulses of radiation at radio wavelengths. The most famous supernova remnant is the Crab Nebula in Taurus (the Bull), which is the debris of a supernova seen in 1054 (though since the Crab is 6,000 light-years away, the actual outburst happened in prehistoric times).

The Crab supernova became visible with the naked eye in broad daylight, and remained visible for several months. But the 1006 supernova in Lupus was much brighter still. Unfortunately we do not know a great deal about it. We have to depend upon fragmentary reports, mainly by Chinese observers—and from any part of China, Lupus is very low in the sky. Apparently the star was visible for two years before fading below naked-eye visibility, and its apparent magnitude was probably about − 10, comparable with the half-moon. Presumably it would have cast strong shadows if it rose to any reasonable height above the horizon.

Many old supernovæ can be traced today because radio waves can still be picked up from their sites. The Crab Nebula is one of the strongest radio sources in the sky. The Lupus supernova is not marked by anything of comparable power, but there does seem to be a good chance that it can be identified with a radio source known by its catalogue number of G.327.6 + 14.5. Its distance seems to be no more than 3,000 light-years.

Since then there have been only three supernovæ seen in our galaxy: the Crab star of 1054, the star of 1572 (Tycho's Star) and that of 1604 (Kepler's Star), though it seems likely that a supernova occurred in Cassiopeia around 1702 and was missed because its glory was hidden by intervening 'dust'. Since the invention of the telescope no galactic supernovæ have blazed forth, and we have to rely upon what we can find out from supernovæ in remote galaxies millions of light-years away. Modern astronomers would dearly like the chance of studying a supernova from closer range, but it is hardly likely that we will be fortunate enough to see an outburst as spectacular as that of 1006 in the Wolf.

31 NUMBER 19 NEW KING STREET, BATH

Go to New King Street in the historic city of Bath, and you will find rows of outwardly undistinguished terraced houses. It would be idle to pretend that the street is one of the more beautiful parts of the city, but one of its houses, Number 19, has an honoured place in the story of astronomy. It was from the garden of

William Herschel (Wedgwood).

Number 19, in March 1781, that William Herschel discovered the planet Uranus.

At that time Herschel was still a professional musician, and astronomy was merely a hobby, but he had taught himself how to make reflecting telescopes (which were, incidentally, the best of their time), and helped by his sister Caroline he had set out to 'review the heavens', mainly with the object of finding out how the stars were distributed in space. The Herschels had lived at Number 19 for two years from September 1777; they had then moved away to another part of Bath, but in March 1781 they decided to return to Number 19. William took up residence first. Never willing to sacrifice a night's observation, he erected his telescope in the garden and began work. It was then he identified the unusual object which he first took for a comet, but which proved to be a new planet.

Almost overnight he became famous. King George III of England and Hanover appointed him King's Astronomer, and before long the Herschels had to move closer to the royal residence at Windsor; they went first to Datchet and then to Slough. Neither residence still exists. The Datchet mansion was ruinous even when the Herschels were there, and Observatory House, Slough, was pulled down in 1960 (its site is now occupied by the Rank Xerox building). Therefore, 19 New King Street is the only 'Herschel house' left standing.

I came into the story in 1978, as a result of a letter from Miss Phillipa Savery, who was (and is) enthusiastic about the history of Bath. Number 19 was in bad repair; what could be done to save it? More in hope than in expectation we formed the William Herschel Society; meetings were held, and then, through the generosity of Dr and Mrs Leslie Hilliard, Number 19 was obtained, renovated and turned into a Herschel museum. The authorities at Flamsteed House (the old

7-ft focus Herschel reflectors, identical to the telescope used to discover Uranus.

Royal Observatory at Greenwich) were amazingly helpful; so were the local astronomers in Bath itself, and Number 19 was opened as a small but important Herschel museum, with many original relics and a superb model of the 'Uranus telescope'.

There have been many visitors. At the moment, sadly, the future is uncertain; even a small house such as Number 19 needs money for maintenance, quite apart from incidental expenses such as rent and power bills. We very much hope that the museum can be kept going. If, therefore, you happen to be in Bath, do not fail to go and see it. After all, it honours an astronomer who brought lustre not only to Bath, but to the whole scientific world.

32   TELESCOPIC FAULTS

During the last few months of 1983 the planet Venus was a brilliant object in the eastern sky before dawn. No early riser could fail to notice it. I had many letters about it, and in particular I received one from an amateur enthusiast who had looked at Venus through his small telescope and had seen a smaller object with a similar phase close beside it. Was this, he asked, a satellite of Venus?

The answer was 'No', because Venus has no moon. But I knew at once what the observer had seen. It was a telescopic ghost. With anything as bright as Venus, the smallest imperfection in the telescope is liable to show up, producing one or more false images. Actually, I own an eyepiece which will show a whole range of ghosts round any brilliant object.

There is an easy way to tell whether an image is a ghost or not. Simply rotate the eyepiece. If the object rotates with it, then you will know at once that it is a ghost. But what are other common telescopic faults?

Be suspicious at once if you see a star showing a definite disk, or, in an extreme case, looking like a large, shimmering balloon. No star (apart from the Sun, of course) can show a perceptible disk. It ought to appear as a point; if it does not, then there is something wrong. Nearly always the trouble is due to poor focusing. If the optics are not focused perfectly, the result will be a blurred, enlarged image. The remedy is to rack the eyepiece in or out until the star looks like a dot.

My 15-in reflector at Selsey.

Yet suppose you cannot get rid of the disk? More investigation is needed; if the trouble is not due to poor focusing, it may be caused by faulty lining-up of the optics. This is betrayed by an image which is not symmetrical. For instance, a star may have a 'flare' to one side or the other. The way to check this is to put the star's image deliberately out of focus. A reflector should then show a symmetrical disk with a hole in the middle. If the hole is not in the middle, or the image bulges out to one side, adjust the small flat mirror in the tube until you have achieved a symmetrical image, and this can usually be done.

On the whole, refractors are less temperamental than reflectors, and unless roughly treated will not need adjusting for many years. Reflectors need more attention, and the mirrors have to be periodically re-coated, usually with silver or aluminium, after which the optics have to be carefully lined up once more. If you continue to see a poor image, remember to test with various eyepieces. The eyepiece which you are using may be the cause of the problem.

Obviously a telescope must have a firm mounting, but sometimes there are other hazards which are less easy to trace. Some time ago I found that my 15-in reflector was not performing as well as usual—and I finally discovered that the main mirror was too tight in its cell, so that it was being 'pinched' and distorted. So if you have a problem with your telescope, make every possible check. In nine cases out of ten, you can identify the trouble and cure it.

33   MOONS OF THE MOON?

Everyone has heard of the 'seas' of the Moon. They are visible with the naked eye, and have been given romantic names such as the Mare Serenitatis (Sea of Serenity), Mare Imbrium (Sea of Showers), Oceanus Procellarum (the Ocean of Storms) and so on. Yet they have never contained any water; analyses of the rocks brought home by the Apollo astronauts and the Russian unmanned probes have shown that they have always been completely dry. They are, in fact, old lava-plains, and were certainly fluid well after the bright highlands had solidified. (It is strange that not so long ago it was still believed that they had once been seas in fact as well as in name. I remember a conversation I had with one of the world's great astronomers, the late Professor Harold Urey, at an international meeting in Prague in 1966. Urey was then convinced that the lunar seas had once been water-filled.)

Craters dominate the lunar scene. We are not sure how they were formed; many astronomers (particularly in America) believe that they were produced by meteoritic bombardment, while others prefer to regard them as volcanic. No doubt both types exist. Meanwhile, in 1983 a leading British astronomer, Professor S.K. Runcorn, produced an interesting modification of the impact theory.

Runcorn starts by considering the Moon's magnetic field—or, rather, lack of it; to all intents and purposes the Moon today has no overall magnetic field. We believe that the Earth's magnetic field is due to dynamo action in the liquid iron-rich core. The Moon has no comparable core now, but on Runcorn's theory there used to be a core which was then melted by the presence of very heavy radioactive elements. These elements decayed after only a few hundreds of millions of years,

The craters Copernicus (right) and Eratosthenes (left), photographed by Commander H.R. Hatfield (12-in reflector). Copernicus is 56 miles in diameter. Note the start of the Apennines, leading away to the left of Eratosthenes.

and the lunar core became 'dead'. This means that in its early history the Moon had a definite magnetic field which it has now lost.

By studying the lunar samples, and the observations made from probes, Runcorn finds that there is evidence of 'residual magnetism' in the rocks, and that the directions of this magnetisation are not random. The ages of the rocks are known, and so the shifting of the Moon's former magnetic poles can be deduced. Since these are not likely to differ much from the poles of rotation, we can work out how the lunar equator has shifted. Runcorn comes to the conclusion that around 4,000 million years ago the present major seas, Imbrium and Serenitatis, lay along what was then the Moon's equator. This can hardly be coincidence, and Runcorn suggests that the large lunar seas were produced by the impacts of former 'moons of the Moon' which used to circle above the lunar equator and then crash-landed.

Certainly it is an intriguing theory, and it seems plausible enough. One thing at least is certain: the Moon has no satellite now. If any such satellite existed, it would have been discovered long ago.

34 SPIRIDION GOPCEVIC: 'LEO BRENNER'

Astronomical charlatans are, mercifully, few and far between. The Chevalier D'Angos was undoubtedly one. It has often been said that another was Spiridion Gopcevic, whose astronomical career was carried out under the pseudonym of Leo Brenner. This may be unfair, but certainly he was either a fraud or else unbalanced.

He was born in 1855 at Trieste, then in Austria; his father, a wealthy ship-owner, committed suicide in 1861, and his widow and young son went to Vienna, where Mrs Gopcevic soon died. Little is known about Spiridion's education, but he started to write books about wars in the Balkans (thereby giving the false impression that he had played an active role) and also wrote about politics; he even produced a few novels. He married a rich wife, and then settled in Lussinpiccolo, the main town in the Adriatic island of Lussin. His wife's money, plus a grant from the Austrian Government, enabled him to erect an observatory, which he called the Manora Observatory; it was equipped with a 7-in refractor, together with various other instruments. Gopcevic changed his name to Brenner, and began to observe the Moon and planets.

At first his work was well received, and was widely published in scientific journals, but before long doubts started to creep in. For example, he fixed the rotation period of Venus as 23 hours 57 minutes 36.2396 seconds (actually it is 243 days!) and his maps of Mars were frankly absurd; he drew hundreds of canals, which he was convinced were artifical waterways. Other observations were equally strange. For example, he published measurements of the faint Companion of Sirius at a time when it was quite beyond the reach of his 7-in telescope.

To make matters worse, he never failed to take criticism in the worst possible spirit, and some of his comments were so extreme that they gave the impression of being pathological. Journals ceased to accept his observations. Brenner then founded his own periodical, and continued to publish it until 1909, but by then his career had taken a sharp downward turn. What happened to his wife is not known, but Manora Observatory was sold in 1909, and 'Leo Brenner' disappeared. As Spiridion Gopcevic he went to America and then returned to Austria, but that is as much as we know. He probably died some time between 1910 and 1935.

At least he has one memorial: a low-walled crater on the Moon, near the large walled plain Metius, was named after him by his friend Philipp Fauth, and the name has been retained on modern lunar maps.

Was Brenner a fake, or was he honest but misguided? We will never know now. We remember him only because of his crater on the surface of the Moon.

35 DELTA PAVONIS

Pavo, the Peacock, is not one of the more imposing constellations. It is a member of the group known as the Southern Birds, and it is too far south to be seen from Europe. However, one of its stars, Delta Pavonis, is interesting because it is so very like our own Sun.

Delta Pavonis is not bright. Its magnitude is 3.6, so that it is easily visible with

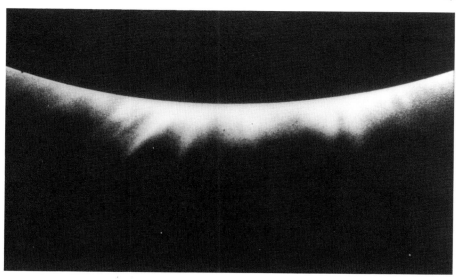

The sun's inner corona. Does Delta Pavonis have a corona of this sort?

the naked eye, but at first glance there is nothing to mark it out. Its distance from us is 19 light-years, which makes it one of our nearest stellar neighbours, and it is slightly yellowish. Its surface temperature is practically the same as that of the Sun, and so is its luminosity. Its mass and diameter are also about the same as those of the Sun. In fact, if you replaced the Sun by Delta Pavonis, the difference would be very slight.

It is believed that the Earth and all the other planets in the Solar System were formed, between 4,500 and 5,000 million years ago, from a cloud of dust and gas or 'solar nebula' associated with the youthful Sun. If the Sun and Delta Pavonis are so alike, why should not a similar planetary system have developed there too?

There seems absolutely no reason why not. But, sadly, we cannot see planets of other stars, because they are too small and faint to show up over immense distances; and if there is an Earth-type planet moving round Delta Pavonis, we have no possible chance of detecting it, even with the Space Telescope which is due to be launched within the next few years. Of course, I am not claiming that there really is a planet associated with Delta Pavonis, earthlike or otherwise. All I am suggesting is that such a planet could well exist. And if there is an 'Earth' moving round Delta Pavonis at a distance of around 93 million miles, there seems no reason why life should not have appeared there. Taking matters a step further, we can suggest that life will be very much on the same lines as our own.

Naturally, this is pure speculation. There may be no planets near Delta Pavonis; even if there are, they may be quite unlike the Earth; and neither do we have any proof that life will appear wherever the environment is suitable for it. Some leading modern astronomers, notably Sir Bernard Lovell, believe that life is a very rare phenomenon in the universe. Yet when we look at Delta Pavonis, we are seeing a sun very similar to ours, and one cannot rule out the possibility that at this very moment some astronomer in the Delta Pavonis system may be looking towards our own world.

36 ANTONIADI AND THE MAPPING OF MERCURY

Before the epic flight of Mariner 10, very little was known about the surface features of Mercury. After all, Mercury is not a great deal larger than the Moon, and it never comes much within 50 million miles of us, besides which it is always to be seen in the same part of the sky as the Sun. Yet ground-based observers did their best to map it. The best effort was due to Eugenios Antoniadi.

Antoniadi was Greek by birth, but emigrated to France and spent most of his life there, finally taking French nationality. For a time he teamed up with another famous French astronomer, Camille Flammarion, but then he was invited to go to the Observatory of Meudon, outside Paris, which is equipped with a 33-in refractor—one of the largest telescopes of its type, and also one of the best (as I know, because I have often used it myself). Antoniadi began to use the 33-in for planetary work. In the 1930s he published a book about Mars, containing his own observations together with a map. The accuracy of this map has proved to be quite remarkable, though, needless to say, Antoniadi saw no Martian craters, and had no means of telling that they existed. (Yes, I appreciate that both E.E. Barnard and John Mellish reported craters, but the telescopes they used were even larger than the Meudon 33-in.)

Mercury was a different proposition. Antoniadi followed the principle adopted by an earlier observer, Giovanni Schiaparelli. If a planet is low in the sky, its light will be coming to us through a thick layer of the Earth's atmosphere, and the image is bound to be violently unsteady. Much better to wait until the planet is high up. With Mercury, this means that the Sun will also be above the horizon, but this was a limitation which Antoniadi accepted.

Using the Meudon refractor, he studied Mercury on every possible occasion, and eventually published a small book about it, together with a map of the surface. I

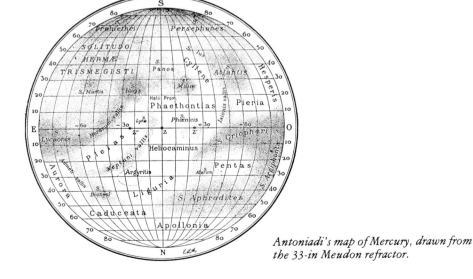

Antoniadi's map of Mercury, drawn from the 33-in Meudon refractor.

first read this book when I was in my teens, around 1935, and a few years ago I translated it into English, but by then it was of historical interest only, because Mariner 10 had already made its fly-by.

Antoniadi named the dusky areas which he saw on Mercury. One patch was christened Solitudo Hermæ Trismegisti, or the Wilderness of Mercury the Thrice Greatest. He also saw lighter areas, and believed that the Mercurian atmosphere was dense enough to support dusty veils which sometimes hid the surface features. But when his map is compared with the actual surface topography as revealed by Mariner 10, we have to admit that there is little resemblance, and no attempt has been made to retain the names which Antoniadi gave to the features he believed he had seen.

This was not Antoniadi's fault. The situation was simply too difficult, and nobody could have done better, particularly since we now know that whenever Mercury is best placed for observation it turns the same face towards us. Antoniadi died in Occupied France in 1942; it is sad that he did not live long enough to see what Mercury is really like.

37 A NEW LOOK AT CANOPUS

Of all the stars in the sky, two stand out because of their exceptional brilliance: Sirius and Canopus. Sirius is visible from Europe, and is a pure white star, though when low over the horizon it seems to twinkle in various colours. Canopus looks almost a magnitude fainter, and is so far south in the sky that it never rises from any part of Europe or the northern United States. Like Sirius, it seems white (though in fact Sirius has a slightly hotter surface), but otherwise the two are as unlike as they could possibly be. Sirius, at a distance of 8.6 light-years, is one of our closest stellar neighbours, and is 'only' 26 times as powerful as the Sun. Canopus, on the other hand, is a real cosmic searchlight.

Measuring star-distances is by no means easy. With the closer stars, out to a few hundred light-years, it can be done by direct observation using the parallax method (remember that a light-year is the distance travelled by a ray of light in one year: almost six million million miles). With more remote stars we have to use indirect methods, and there are bound to be uncertainties. Canopus is a very long way away, and estimates of its distance have varied widely, but according to a very recent measurement it is at least 1,700 light-years from us. We are seeing it as it used to be in the third century AD.

If so, then because Canopus looks so bright, it must be highly luminous. And if this new measurement is correct, then the power of Canopus must be around 200,000 times that of the Sun.

Try to imagine it—200,000 Suns put together! It defeats all our powers of comprehension, though let me add at once that even Canopus is not the holder of the 'luminosity record'. It is certainly beaten by another star in the same constellation, Eta Carinæ, which however is so remote that it cannot be seen at all with the naked eye.

If Canopus is so luminous, then it must be squandering its energy reserves at a furious rate. A mild star such as the Sun can shine steadily for many millions of years, and even though the Sun is losing mass at the rate of 4 million tons every second it will not change much for at least 5,000 million years in the future. But

Canopus will not last for anything like so long as that. It will use up its available 'fuel', and will change drastically. It is so massive that it will suffer a violent outburst of the kind we call a supernova, blowing much of its material away into space and leaving only a small, super-heavy remnant. It may even collapse into what is now known as a black hole.

Of course, this disaster is not imminent. You and I will see no change in Canopus; it has at least a million years' grace, probably longer. But if we could use a time-machine and come back in, say, 1,000 million years, the Sun will be much the same as it is now, while Canopus will not. Like all its kind, it is a cosmic spendthrift.

38 THE POWER OF THE NAKED EYE

How much can one see without optical aid? Obviously, everything depends upon the quality of one's sight, but it is interesting to examine some cases which are at the very limit of visibility.

The Pleiades cluster is a good test. It is known as the 'Seven Sisters', and it is true that people with average sight can see seven individual stars under good conditions; but very keen-eyed people can go much further, and the record is said to be held by a last-century German astronomer, Eduard Heis, who could go up to 19. Then there are the phases of Venus. There seems little doubt that when the planet is in the crescent stage, the phase really can be seen by people with exceptional sight. I once conducted a rather unkind experiment on television. I

Venus as a crescent, photographed by the Palomar 200-in reflector.

showed a photograph of Venus, which at that time was in the crescent stage in the evening sky, and asked viewers to draw the phase—if they could see it—and send the drawings to me. What I had done was to show the telescopic view, which is, of course, reversed. Many people sent in sketches reproducing this, so that obviously they were guilty of 'seeing what they expected to see', but three viewers wrote in baffled vein to the effect that they could see the crescent, but opposite to that I had put on the screen. Those were the genuine sightings.

There is also an interesting fact about the Galilean satellites of Jupiter. But for the overpowering brilliancy of the planet itself, Ganymede at least would be an easy naked-eye object. There are well-authenticated cases that it can be glimpsed, and of these the first goes back to the year 364 BC!

One of the earliest of known Chinese astronomers was Gan De. Nothing is known about his life, and unfortunately his original works have been long since lost, but portions were reproduced in a book on astrology written by Qutan Xida around AD 720. Gan De is quoted as follows:

'In the year of chan yan. . . Jupiter rose in the morning and went under in the evening together with the Lunar Mansions Xunu, Xu and Wei. It was very large and bright. Apparently there was a small reddish star appended to its side. This is called 'an alliance' ('tong meng').'

Tong meng was a term used to describe a close alliance. And since the year can be checked, there seems little doubt that what Gan De saw was a satellite. It must have been either Ganymede or Callisto, which are the furthest from Jupiter, but Callisto is much the fainter, and therefore Ganymede is much the more probable. If so, Gan De anticipated Galileo by almost 2,000 years.

39   SIR JAMES JEANS

There have been many popularisers of astronomy. Undoubtedly one of the best was Sir James Hopwood Jeans, who was born on March 24 1877, in London, and was educated at Cambridge. He first became known because of his researches into the origin of the Solar System; he believed that the Earth and the other planets were pulled off the Sun by the gravitational action of a passing star, and this so-called Tidal Theory was widely accepted for many years, even though it is now known to be wrong. Today, it seems certain that the planets condensed out of a cloud of dust and gas which surrounded the youthful Sun.

Jeans paid great attention to problems of the internal constitution of the stars, and his name is often coupled with that of another famous astrophysicist, Sir Arthur Eddington, though actually their points of view were rather different. Jeans wrote several technical books and many papers, and he was responsible for major advances in our understanding of how the stars evolve. Later in his career he concentrated more and more upon popular writing, and some of his books, such as *The Mysterious Universe* and *The Universe Around Us*, were classics of their time. He had a delightfully fluent style, and a knack of making difficult problems sound easy, though as a true scientist he was always wary of over-simplification.

Jeans was also the first famous radio broadcaster on astronomy. During the 1930s his voice became very familiar to listeners to the BBC; he continued to broadcast regularly during the war, and indeed until a very short time before his death in 1946. It is a great pity that he died before television became common; he would have been so good at it.

Astronomy did not occupy his whole time. He was a man of varied interests, and in particular he was a good musician (his wife, still happily with us, is a famous organist).

Unquestionably Jeans was responsible for encouraging many beginners to take up astronomy as a serious hobby, and even as a profession. He was always ready to give a helping hand. Many present-day scientists owe much to him, and his death left a gap which could not easily be filled. One phrase of his always sticks in my mind. A reporter asked him how he would describe himself. Jeans thought for a moment, and then gave a perfect answer. 'I am,' he said, 'a publicity man for the planets.' How true that was!

40 'THE OLD MOON IN THE NEW MOON'S ARMS'

Many people believe that the thin crescent moon, appearing in the evening sky soon after sunset, is the true 'new moon'. In fact that is not true. New moon occurs when the Moon lies between the Earth and the Sun. Its dark side is then turned toward us, and we cannot see the Moon at all, though if the alignment is perfect we are treated to the spectacle of a solar eclipse.

During the crescent stage, you will probably see the 'night' part of the Moon shining faintly. This is the appearance which country folk often call 'the Old Moon in the New Moon's arms'. Originally it caused a great deal of bewilderment, and it was even suggested that the Moon itself might be faintly self-luminous. However, Leonardo da Vinci, the great Italian scientist and painter who lived between 1452 and 1519, realised the truth. He wrote: 'The Moon has no light of itself, but so much of it as the Sun sees, it illuminates. Of this illuminated part we see as much as faces us. And its night receives as much brightness as our waters lend it as they reflect on it the image of the Sun, which is mirrored in all those waters that face the Sun and the Moon.' In other words, the luminous glow is caused by light reflected on to the Moon by the Earth; nowadays we call it the Earthshine.

It can be quite bright, and with binoculars or a telescope you can see details in the earthlit portion. There is one crater, the 23-mile Aristarchus, which is very reflective by lunar standards, and so is often visible when lit only by earthshine. The great astronomer William Herschel seems to have mistaken the earthlit Aristarchus for an erupting volcano in 1789, and one can hardly blame him.

The earthshine is usually seen whenever the Moon is a slender crescent, though of course a clear sky is needed (mist will drown the faint glow). The fact that the earthshine is not always equally bright has nothing to do with the Moon, but depends upon the state of our own atmosphere. If the Earth is largely cloud-covered, it will reflect more light, and this brightens the earthshine. It has been said that colour can be seen in the earthlit part of the Moon; I have never seen anything of the kind, but positive reports are quite common.

Photographing the earthshine is quite a challenge, because one tends to over-expose the sunlit crescent; but it can be done, and although such photographs are of no scientific value, it is worth taking a picture of 'the Old Moon in the New Moon's arms' for its beauty alone.

41   THE ELUSIVE COMET

Comets are insubstantial, ghostly things, with very little mass by planetary standards. The many short-period comets are predictable; the others have periods so long that we never know when or where to expect them. But there was one interesting case of a comet which was seen on a single occasion only, and which remains very much of a mystery.

The man responsible was Ernst Klinkerfues, a German astronomer of the last century who became director of the observatory at Göttingen. His main work was in connection with the stars, but he was also interested in comets. In particular, he made a careful study of Biela's Comet.

Biela's Comet used to have a period of 6¾ years, and though never bright enough to be seen with the naked eye it was a conspicuous telescopic object. It was seen regularly; it came back in 1826, 1832 and 1845, though it was missed at the 1839 return because it was so badly placed in the sky. In 1845 it astonished astronomers by splitting in half. The twins returned on schedule in 1852, were missed in 1858 because they were again badly placed, and were confidently expected to return once more in 1866. However, they failed to do so. Despite careful searches, they could not be found. In fact they have never been seen again, though in 1872 a major meteor shower came from the position where the comet

*Biela's Comet, which split in half in 1845 (*drawing by A. Secchi*).

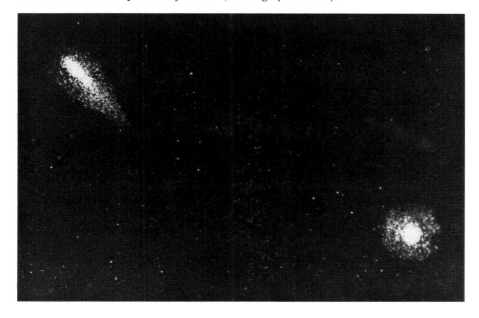

The 1947 eclipse comet. Tewfik's Comet of 1882 was much the same, but the 1947 comet was subsequently tracked.

ought to have been. Some meteors are seen every 6¾ years from the same point, though recently the shower has become very feeble.

Ernst Klinkerfues was intrigued by the whole problem. On November 30 1872 he sent a telegram to a well-known British astronomer, Pogson, who was at Madras in India: 'Biela touched Earth on 27th. Search near Theta Centauri.' Pogson duly looked around the region of the bright star Theta Centauri—and there was a comet. He observed it on December 2 and 3, but bad weather then closed in, and nobody else managed to see the comet at all.

The strange thing about it is that the comet cannot have been Biela's, because the position was all wrong. So what did Pogson see? He was too experienced to report a comet where none existed. And what did Klinkerfues really mean in his enigmatical telegram? Despite a careful search through the literature I have been unable to find any clues. Klinkerfues died in 1883, so we are hardly likely to find out the truth now.

Another comet seen once only was that of 1882, when it was photographed near the totally-eclipsed Sun. The picture was taken from Egypt, and the comet was named Tewfik's Comet after the then ruler of that country, but nothing else is known about it—and never will be.

42   C.A.F. PETERS

Few people today remember the last-century German astronomer Christian August Friedrich Peters. Yet he was a famous scientist, and in particular he was concerned with the distances of the stars.

Peters was born in Hamburg in 1806, and studied at the University of Königsberg. His tutor there was Friedrich Bessel—and it was Bessel who made the first measurement of the distance of a star. His method was that of what we now term parallax.

To show what is meant, a simple experiment will suffice. Close one eye, hold out a finger at arm's length, and line it up with something in the background (a picture on the wall will do quite well). Now, without moving your finger or your head, shut the other eye and open the first. You will see that your finger is no longer lined up with the picture. Your two eyes are not in exactly the same place, so that you are observing from a slightly different direction. The apparent shift of your finger against the background picture is a measure of its parallax.

What Bessel did was to use the movement of the Earth. We are 93 million miles from the Sun, so that the full diameter of the Earth's orbit is 186 million miles. Bessel selected a star which he thought might be nearby (a faint dwarf in the Swan, 61 Cygni), and measured its position with respect to the more remote stars in the background. Six months later, when the Earth had moved round to the other side of its orbit, he repeated the measurement, and found a definite parallactic shift. Going back to our experiment: the finger represents the nearby star, the picture represents the remote background stars, and your two eyes represent the Earth at a six months' interval.

Having measured the parallax angle, and knowing the diameter of the baseline (186 million miles), Bessel was able to calculate the distance of the star. It turned out to be about 11 light-years. About the same time, Henderson in South Africa made similar measurements for the brilliant star Alpha Centauri, and arrived at the correct answer: Alpha Centauri is 4.2 light-years away.

By the time that Bessel made his announcement, in 1838, Christian Peters had moved on to Altona to become director of the observatory there. In 1878 he went to Kiel, again as director of the observatory. He continued the work on stellar

The principle of parallax.

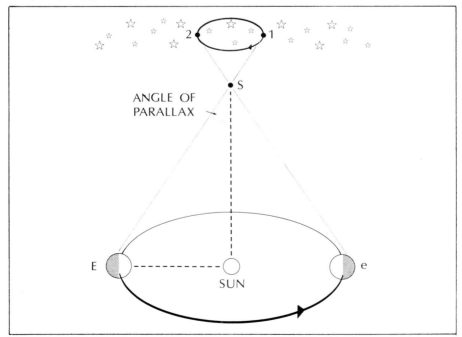

parallaxes, and measured a great number of them successfully before his sudden death in May 1880.

When talking about the history of measuring star distances, everyone remembers Bessel and Henderson. But let us not forget Peters, who made a major contribution, and who has an honoured place in the story of astronomy.

43   LIFE FROM SPACE?

In Victorian times there was a tremendous furore when Charles Darwin (and let us not forget Alfred Russel Wallace) put forward a theory of evolution, according to which *homo sapiens* evolved from small tree-living primates. People jumped to the conclusion that according to Darwin, men were descended from monkeys. In fact Darwin never said anything of the kind; all he maintained was that men and monkeys have a common ancestry. But whether this was so or not, few people have doubted that life on Earth began on Earth.

One doubter was the Swedish Nobel Prize winner Svante Arrhenius, who proposed a 'panspermia' theory according to which life was brought to the Earth by way of a meteorite. The theory never became popular, but recently something of the same idea has been popularised by two very eminent astronomers, Professor Sir Fred Hoyle and Professor Chandra Wickramasinghe, who claim that comets are the main agents in moving organic matter round the universe. They also maintain that cosmic grains, spread between the stars, are nothing more nor less than bacteria.

According to Hoyle and Wickramasinghe, life had to begin in space because it required all the vast extent of the universe to make it possible. The initial building-up of a living organism requires a whole chain of events, each of which is in itself most unlikely; and cosmically speaking, the Earth is a very small place. The reasoning is, therefore, that life originated far away from us, and took hold here simply because conditions were suitable for it to gain a foothold. This would not be the case on, say, an airless world such as the Moon. Living material deposited there would be quite unable to survive.

It seems overwhelmingly probable that many stars have planetary systems, and if Hoyle is right we may assume that any planet which *can* support life *will* support life. Moreover, that life will develop as far as its environment allows. In the Solar System, unfortunately, there seem to be few opportunities; most of the planets are airless, lack solid surfaces, or are too hot or too cold. The only possible candidate seems to be Mars.

The Viking landers have shown no traces of Martian life, but this does not disprove Hoyle's theory, because conditions in the past may well have been much more favourable than they are today. It is at least possible that life began there and became extinct when conditions worsened. One way to find out is to analyse Martian material, and see whether it contains any positive evidence in the form of what we may loosely term fossils. Short of a manned expedition, this means using a there-and-back automatic probe, as the Russians have already done for the Moon. This is certainly coming within the range of technical possibility, and I have little doubt that it will be achieved before the end of the century.

Once Martian samples can be studied, we will have much more information to guide us. If traces of past or present life are found, we will be forced to conclude that life will appear wherever it can find suitable conditions. If there are no signs of Martian organisms, the Hoyle-Wickramasinghe theory will be weakened, though not necessarily abandoned; Mars could be a special case.

To my mind, this is one of the most important of all the investigations to be carried out in the near future. Remember that though most astronomers believe that life is likely to be common, there are some eminent dissentients, and we cannot say for certain that we are not alone in the universe.

I was present when the original lunar samples from Apollo 11 were first opened, in 1969. I would dearly like to be present also when the first samples from Mars are examined. They could well clear up the most fascinating of all the problems facing us today.

44　THUBAN

One of the 48 constellations listed by Ptolemy around the year AD 150 is Draco, the Dragon. In mythology it seems to represent either the dragon killed by the hero Cadmus before the founding of the city of Bœotia, or else the dragon which guarded the golden apples in the garden of the Hesperides. Draco is a long, sprawling constellation, winding its way between the two Bears and ending up not far from the brilliant blue Vega, in Lyra. It is so near the north pole of the sky that it never sets over England, but it is not striking. The brightest star, Eltamin in the dragon's head, is below the second magnitude.

However, there is one star—Thuban, or Alpha Draconis—which is worthy of note. Its magnitude is 3.6, so that it is rather fainter than Megrez, the dimmest of the seven stars which make up the pattern of the Great Bear or Plough. It is white, with a surface hotter than that of the Sun; it has about 150 times the Sun's luminosity; and it is approximately 230 light-years away, so that we are now seeing it as it used to be in the time of King George II. It is not hard to identify, because it lies midway between Alkaid, the end star in the Great Bear's tail, and Kocab, the 'Guardian of the Pole' in Ursa Minor, the Little Bear.

What, then, makes Thuban interesting? Simply that in ancient times, when the Egyptian Pyramids were built, it used to be the north pole star.

The north celestial pole is that point in the sky which lies in the direction of the axis of rotation—so that if you go to the North Pole of the Earth, the celestial pole will be directly overhead. But the position of the pole changes. The Earth is not a perfect sphere; it bulges slightly at its equator, and the Sun, Moon and planets pull upon this bulge, causing the axis to 'wobble' rather in the manner of a gyroscope which is running down and is starting to topple. Naturally, this shifts the position of the poles of the sky. It takes the Earth's axis about 26,000 years to describe a full circle, so that the change is very slow, but over the centuries it mounts up. In the days of Ancient Egypt, the pole lay close to Thuban.

Today, of course, it is within one degree of Polaris in the Little Bear. By AD 4000 it will have moved into the constellation of Cepheus; by AD 10000 it will have reached Cygnus; and in the year AD 14000 it will be near Vega. It will then move back into Draco, before returning to the neighbourhood of Polaris 26,000 years

Star trails. The North Celestrial Pole is in the centre. Note the short trail left by Polaris, within 1° of the pole. Thuban is now well away from the polar point.

hence. This is the phenomenon of 'precession'; it was discovered by the Greeks well before the time of Christ.

So I suggest that you go and find Thuban next time the sky is dark and clear. Unspectacular though it is, there will come a time, in the far future, when it will again be the north polar star.

45 COSMIC DISASTERS

Not long ago an interesting study was produced by students at the Massachusetts Institute of Technology concerning Icarus. Icarus is an unusual minor planet or asteroid; its orbit takes it within the orbit of Mercury, and at perihelion it is a mere 16 million miles from the Sun, so that it must be red-hot. It remained the record holder until 1983, when the IRAS (Infra-Red Astronomical Satellite) detected another tiny body, 1983 TB, which moves to within nine million miles of the solar surface.

Icarus is in no danger of colliding with the Earth at the current epoch, but in the future it could conceivably do so. The MIT study described the various counter-measures that might be taken if we found that Icarus (or a similar body, a mile or two in diameter) was on a collision course.

Trail of Icarus, photographed by R.S. Richardson. Icarus is the short streak. It moved while the time-exposure was being made (Mount Wilson Observatory).

Obviously there are only two possibilities: either destroy the asteroid by nuclear explosion, or deflect its orbit and make it miss the Earth. Neither procedure is easy, but, given enough time, it could be carried out. Unfortunately there might not be enough time, and Icarus-sized asteroids seem to be much more common than used to be thought (according to one theory they are simply the nuclei of dead comets which have lost all their gas). However, it must be added that the Earth must be regarded as a small target, and a disastrous asteroid collision must be a rare event.

What, then, about a cataclysmic 'close encounter' with another star? Here the danger seems very slight indeed. Apart from the Sun, the nearest stars are those of the triple Alpha Centauri system, at over four light-years; next comes a dim red dwarf, Barnard's Star, at six light-years. Were there any stars closer than that, I think it is fairly certain that they would have been found by now. (Some years ago there was a flurry of excitement when O.J. Eggen, from Australia, announced the detection of a faint star which he believed to be closer than Alpha Centauri, but this proved to be an error; Eggen's Star does not even come into the list of the closest dozen.) There is no chance whatsoever of a luminous star threatening us in the foreseeable future, and more probably it will never happen.

We must also consider dark, dead stars which have used up all their energy, and are not radiating at any wavelength. It seems that stars of about the Sun's mass will end up in this way, as 'black dwarfs', after having spent thousands of millions of years as feeble white dwarfs, but it is not certain that the universe is old enough for any black dwarfs to have developed yet.

However, if a stellar corpse did invade the Solar System, we would know of its existence long before we could actually see it, because its mass would be comparable with that of the Sun, and it would cause marked perturbations in the orbits of the planets—first the outer ones, then the inner. The damage it would do would be incalculable even if there were no direct collision, because even minor changes in the Earth's path would alter the climates so markedly that life could well be wiped out. There would be nothing that we could do about it; we would simply have to wait and hope for the best. Luckily, the approach of a dead star is most improbable. I am merely pointing out that the chances of such a catastrophe are not absolutely nil.

46 PER WARGENTIN AND THE VANISHING MOON

Unless you are interested in the history of science, it is not likely that you will have heard anything about Per Vilhelm Wargentin. Nevertheless, he was an 18th century Swedish astronomer of considerable eminence, and he was associated with several interesting events.

Wargentin was born on September 22 1717, and after taking his degree at the University of Upsala he went to Stockholm, to become director of the observatory there. At an early age he had been attracted to astronomy by an eclipse of the Moon, and it was this which launched him upon his career.

A lunar eclipse occurs when the Moon passes into the cone of shadow cast by the Earth. Its direct supply of sunlight is temporarily cut off, and the Moon turns a dim, often coppery colour until it emerges from the shadow. Generally it does not vanish completely, because some of the Sun's rays are bent on to the lunar surface by way of the blanket of atmosphere surrounding the Earth. But on May 18 1761 things were different. Wargentin wrote: 'The Moon's body disappeared so completely that not the slightest trace of the lunar disk could be seen either with the naked eye or with the telescope, although the sky was clear, and stars in the vicinity of the Moon were distinctly visible in the telescope.' Since any sunlight reaching the eclipsed Moon must first pass through the atmosphere of the Earth, conditions in our upper air are all-important. Two years earlier, in 1759, the Mexican volcano Jorullo had erupted violently and hurled large quantities of dust and ash into the top part of the atmosphere, and this took a long time to settle, which no doubt accounts for Wargentin's observation.

Other dark eclipses have been those of December 1601, June 1620 and October 4 1884, but the last occasion upon which the Moon completely vanished was that of the eclipse of June 10 1816. Once again a major volcanic outbreak was responsible. In April 1815 the volcano Tambora, on the Indonesian island of Sumbawa, exploded so catastrophically that over 90,000 people were killed, either directly or by subsequent starvation as the crops failed. Indeed, 1816 has been termed 'the year without a summer'; it was even worse than that of 1815 when the actual eruption occurred. On the other hand the eclipse of March 19 1848 was so

Lunar eclipse photographed by Commander H.R. Hatfield.

bright that the Moon turned blood-red, and many people doubted whether an eclipse was in progress at all.

Per Wargentin himself carried out much valuable work; he published excellent tables of the movements of Jupiter's four bright satellites, and did much to organise astronomy in Sweden. He died in December 1783. It is fitting that his name should have been attached to a formation on the Moon—a 55-mile plateau or lava-filled crater, unlike anything else on the lunar surface. Certainly Wargentin deserves to be remembered.

47 HOW WRONG WE WERE!

In the summer of 1964 I was invited to give a lecture at Cambridge University. My subject was 'Mars'. It was well attended, and I gave a general summary of what we knew, or thought we knew, about the planet. I made a series of 12 profound statements, each of which was backed up by the best available scientific evidence—and every one of which turned out to be wrong.

This was only two decades ago, but it was before the first successful Mars probe, Mariner 4, made its fly-by and turned all our pre-conceived ideas upside down. Before Mariner 4 we had been confident that Mars had a flattish or at most a gently undulating surface; that the atmosphere was composed mainly of nitrogen, with a ground pressure of about 85 millibars; that the white polar caps were due to a wafer-thin layer of hoar-frost; and that the dark areas were old sea-beds filled with a primitive organic material ('vegetation', if you like). Though nobody retained much faith in the age-old idea of intelligent Martians, it was still believed that the 'canals' had a basis of reality.

The probes, beginning with Mariner 4 in 1965, have shown otherwise. The atmosphere is chiefly carbon dioxide, with a ground pressure of below ten millibars everywhere. The dark regions are not always depressions, and are not due to living organisms; they are merely what may be termed 'albedo features', differing from their surroundings only by their darker hue. The residual polar caps are made of ordinary ice, and are very thick, though in Martian winter they are coated with a thin layer of carbon dioxide ice. The canals do not exist in any form, and the spider's-web network of Lowell and others was purely imaginary. And finally, the surface is very uneven; one volcano, Olympus Mons, towers to 15 miles above the ground below. We were wrong even about the colour of the Martian sky. We had expected it to be dark blue; in fact it is salmon-pink.

Why were we so wrong? The reason, of course, is that we had had to observe Mars across a distance of at least 34 million miles, so that no telescope would show it more clearly than the Moon appears when viewed through low-powered opera glasses. For that matter, there had been some strange ideas even about the Moon.

Left *Venera 1, the first (unsuccessful) Russian probe to Venus.*

Above right *The south polar cap regions of Mars in late summer, as photographed from Viking Orbiter 2. The south pole is just off the lower right edge of the bright residual cap. The area shown is 800 miles across.*

Right *Mariner 9, the first spacecraft to send back really good photos of Mars showing the giant volcanoes.*

Above *Utopia: the Martian landscape from Viking 2.*

Left *Launch of Mariner 9, on May 30 1971, from Cape Canaveral. It took 167 days to reach the neighbourhood of Mars.*

Above right *Computer-enhanced mosaic of Mariner 9 photos showing the volcano Olympus Mons; the Viking high-resolution photos show details not evident in this one, producing new geologic interpretations of Martian history.*

Right *Mosaic of photos from Viking Orbiter 1 showing the area marked in the previous illustration in far greater detail. The summit caldera is a complex feature recording a series of eruptions in varied levels of frozen lava lakes.*

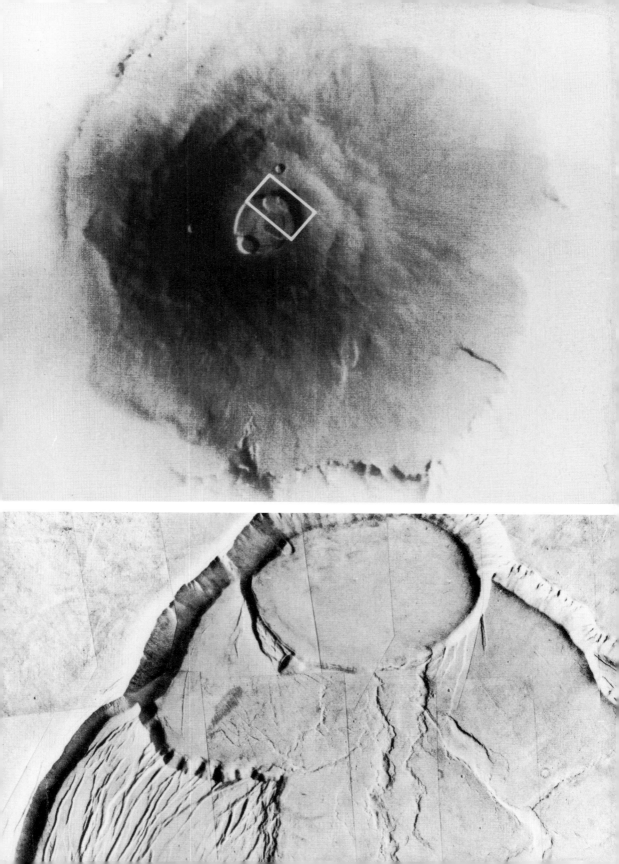

Venus: the surface as seen from Veneras 9 and 10, June 1975.

One of the most peculiar was due to Dr Thomas Gold, then of Cambridge University, in the 1950s. Gold assumed the lunar 'seas' to be deep dust-drifts, so that any space-craft unwise enough to land would, in Gold's words, 'simply sink into the dust with all its gear'. Few practical lunar observers agreed (I for one had no faith in it), but it was taken very seriously by the American authorities, and was not finally destroyed until the Russian automatic probe Luna 9 made a controlled landing and showed no inclination to sink out of sight.

We were equally wrong with Venus. Before 1962, much was heard of a theory due to D.H. Menzel and F.L. Whipple, according to which the Cytherean surface was covered mainly with water, with surrounding clouds of H_2O. It was known that the atmosphere is rich in carbon dioxide, so presumably the water would have been fouled, changing it into soda-water. In this case, life might well have started there, just as it did in the warm seas of Earth so long ago. We know better now. The surface temperature is over 900 degrees Fahrenheit, so that liquid water cannot exist, and the clouds contain large quantities of sulphuric acid. Instead of being more welcoming than Mars, Venus has turned out to be probably the most hostile of all the planets in the Solar System.

The Red Spot on Jupiter provided another instance. It was widely believed to be a solid or semi-solid body floating in the Jovian gas; in fact it is a whirling storm—a phenomenon of Jupiter's 'weather'. And the rings of Saturn, once thought to be more or less uniform with only one major gap (Cassini's Division) have turned out to be incredibly complex, with thousands of thin ringlets and narrow gaps.

In stellar astronomy there have been equally startling revisions of thought. Red giants such as Antares, once assumed to be youthful, are now known to be well advanced in their life stories. And as recently as 1920 Dr Harlow Shapley, the man who first measured the size of the Galaxy, was still defending his opinion that the objects then termed spiral nebulæ were contained in our own Galaxy rather than being independent systems.

The point is, surely, that all these mistakes—and many more of the same kind—were based on the best available evidence. I have a feeling that by, say, AD 2000 we will find that many ideas current in 1984 are equally wide of the mark. Time will tell.

48 JOHANN SCHRÖTER AND THE SCHRÖTER EFFECT

Have you ever heard of Johann Hieronymus Schröter? Unless you are astronomically-minded, probably not. He was a German amateur astronomer who lived at Lilienthal, near Bremen. I never met him—which is hardly surprising, because he died in 1816. But I have a great admiration for him, and historians of science tend to underrate him badly.

Schröter was Chief Magistrate of Lilienthal, and to him astronomy was a hobby. He set up an observatory, and equipped it with powerful telescopes. The largest of these was a 19-in reflector by an instrument maker named Schräder, about whom not much is known expect that he lived in Kiel and was very deaf. The quality of the 19-in is uncertain, but Schröter carried out most of his work with smaller telescopes, at least one of which was made by no less a person than William Herschel.

Schröter was concerned mainly with the Moon and planets. He made thousands of lunar drawings, and though he was not a skilful draughtsman he was certainly accurate, so that he ranks as the first really great selenographer or Moon-mapper. He made sketches of Mars which were better than any previously produced, and he also studied Venus, on which he could see few markings (well, who can?), but whose phase was of considerable interest to him.

Venus, closer to the Sun than we are, shows phases similar to those of the Moon, because, obviously, the Sun can shine on only half the planet at any one time, and all depends upon how much of the sunlit half is turned in our direction. It should be easy to calculate just when Venus will be at exact half-phase, or dichotomy. However, Schröter found that theory and observation did not agree. When Venus was an evening object, shrinking to a crescent, dichotomy was always early. When Venus was a morning object, and changing from a crescent into a half, dichotomy was late. Years ago, in a book I wrote about Venus (long before the space age), I christened this phenomenon the 'Schröter Effect', a term which has now been officially accepted.

The discrepancies do exist, and may amount to several days. Undoubtedly the thick atmosphere of Venus is responsible for them, since there is no chance that the calculations are wrong. The precise cause is not known even yet, and the phenomenon is interesting even if it is not of tremendous importance.

Schröter made many other valuable observations. Sadly, his observatory was destroyed by the French during the wars of the early 19th century; all his unpublished manuscripts were burned, and even his brass telescopes were plundered, because the French soldiers believed them to be gold. Luckily, however, enough of his work remains to show what a painstaking and honest observer he was, and he will not easily be forgotten, even though he died so long ago.

49   THE STARS FROM THE BOTTOM OF A WELL

There used to be a comic song which began: 'Where do all the flies go in the winter-time?' More scientifically, I have often been asked: 'Where do all the stars go in the day-time?' The answer is, of course, that they do not go anywhere. They are just as bright as they are during the night, but the sunlit sky effectively drowns them.

The brightest thing in the sky, apart from the Sun and the Moon, is the planet Venus. People with keen eyes can see it well before sunset or after sunrise, and I do know exceptional persons who can pick it out most of the time. I remember, too, the total solar eclipse of June 1983, which I saw from Tanjung Kodok on the island of Java. Venus was visible at least a quarter of an hour before totality began (I did not see it then myself, because I was too busy setting up my equipment) and it was followed for about half an hour after totality had ended. This, of course, was with the naked eye. Telescopically Venus can always be found in daytime when it is not too near the Sun; indeed, this is the best way to observe Venus—and the same is true for the other inner planet, Mercury.

If you have a telescope fitted with setting circles so that you can point it accurately, you can locate bright stars easily in the daytime (though always take great care not to search round in the region of the Sun, and in any case always sweep *away* from the Sun so that there is no danger of looking at it by mistake). In twilight, first-magnitude stars are easy enough. But what about the old story that stars can be seen in full daylight if the observer goes down to the bottom of a deep well or mine-shaft?

I have heard this story many times, but a little thought will show that there is nothing in it. What matters is the *contrast* between the star and the sky background, which is just the same from a well-bottom as it is from ground level. If you doubt me, try for yourself. I have done so—from the bottom of Homestake Gold Mine in South Dakota, a mile below the surface. When I looked straight up the shaft, I could see a circle of blue with no stars. So I fear that another charming old tale has to be disregarded.

Yet what about the airless Moon? The sky there is black all the time. I asked the 'last man on the Moon', Commander Eugene Cernan of Apollo 17, whether he saw stars in the daytime. He replied that he had, though only after having 'dark-adapted' and by shielding his eyes from the glare of the surrounding sunlit rocks. From a space-craft, of course, the stars are glorious and unwinking. But I'm afraid that a gold-mine or a well will not help!

50   WILLIAM HERSCHEL AND THE VOLCANOES OF THE MOON

In the year 1787 William Herschel, by then recognised as probably the world's leading astronomical observer, presented a remarkable report to the Royal Society. He said that without the slightest doubt he had observed active volcanoes on the surface of the Moon. His report read as follows:

'April 19 1787, 10h 36m. I perceive three volcanoes in different places of the

The Oceanus Procellarum (Ocean of Storms) photographed by Commander Hatfield (12-in reflector). Aristarchus, the brilliant crater which Herschel mistook for a volcano, is at the bottom. The ray-crater near the centre is Kepler, and Gassendi is at the top.

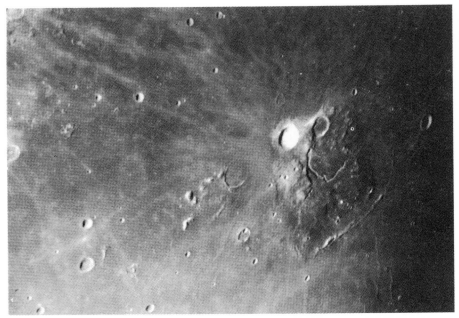

Another photo by Commander Hatfield showing Aristarchus, the dark-floored Herodotus and the magnificent winding valley.

dark part of the Moon. Two of them are either nearly already extinct, or otherwise in a state of going to break out. . . . The third shows an actual eruption of fire, or luminous matter. . . . Its light is much brighter than the nucleus of the comet which M. Méchain discovered at Paris on the 10th of this month.

'April 20, 10.02. The volcano burns with greater violence than last night. I believe its diameter cannot be less than 3 seconds of arc. . . . Hence we may compute that the sunshine or burning matter must be above 3 miles in diameter. It is of an irregular round figure, and very sharply defined on the edges. The other two volcanoes are much farther toward the centre of the Moon, and resemble large, pretty faint nebulæ, that are gradually much brighter in the middle; but no well-defined luminous spot can be seen in them. . . .

'The appearance of what I have called the actual fire or eruption of a volcano, exactly resembled a small piece of burning charcoal, when it is covered by a very thin coat of white ashes, which frequently adhere to it when it has been for some time ignited; and it had a degree of brightness about as strong as that with which a coal would be seen to glow in faint daylight.

'All the adjacent parts of the volcanic mountain seemed to be faintly illuminated by the eruption, and were gradually more obscure as they lay at a greater distance from the crater.

'The eruption resembled much that which I saw on the 4th of May 1783; . . .it differed, however, considerably in magnitude and brightness; for the volcano of the year 1783, though much brighter than that which is now burning, was not nearly so large. . . . The former seen in a telescope resembled a star of the 4th magnitude as it appears to the natural eye; this, on the contrary, shows a visible disk of luminous matter, very different from the sparkling brilliance of starlight.'

Such was Herschel's report. It is, of course, clear that what he saw was not an eruption; all the lunar volcanoes became extinct long ago. There is no doubt that the main 'volcano' was the brilliant crater Aristarchus illuminated by earthlight; Herschel's observations were made when the Moon was a slender crescent (three days past new), and Aristarchus can generally be well seen under such conditions. The other 'volcanoes' must have been of the same nature. There are many craters, such as Dionysius, Censorinus and even Copernicus, which are often prominent by earthshine. It seems that Herschel may well have realised his mistake. He did not refer to the volcanoes again, and this was, indeed, his last published paper concerning the Moon.

51   CANOPSUS STREET

Many of the most interesting objects in the sky lie so far south of the celestial equator that they can never be seen from Europe or most of the United States. This is why there has been recent concentration upon observatories in the southern hemisphere. Parts of Australia provide good observing conditions; so do parts of South America, and in fact three giant telescopes are now sited in Chile. There is also South Africa, which has always been in the forefront of astronomical research. Major observatories were set up at Cape Town, Johannesburg and Bloemfontein. There was also the Radcliffe Observatory, just outside Pretoria. Construction began just before the war, though the main telescope, a 74-in reflector, was not completed until 1948.

Sadly, the Radcliffe Observatory no longer exists. In 1972 the official decision was made to shift most of the big South African telescopes to a new site at Sutherland, in Cape Province, where the seeing is excellent. It was probably a wise decision, and so far as the Radcliffe was concerned it was inevitable, because Pretoria had grown so rapidly that observing conditions had deteriorated. So the observatory was dismantled, and the 74-in is now in full working order at Sutherland. I remember going to the old Radcliffe site in 1975. It looked broken and ghostly. Weeds covered the once-trim paths and gardens, the remains of the dome were in a sorry state, and the whole atmosphere was one of desolation.

Sentiment has no real place in science, though there were many astronomers who regretted the demise of the Radcliffe. But there was one amusing episode which deserves to be put on record.

The observatory was five miles from the centre of Pretoria, at a height of 5,000 ft. The road from the city was long, attractive and wooded (as it still is). It also served as a residential area, and when the Radcliffe was established the Pretoria City Council had the idea of naming local roads after stars and constellations. Orion Street, Perseus Street and the rest remain. One road was named Canopsus Street—Canopsus, note, not Canopus. This was an error by the Council official who allotted the names and put up the road-signs.

For a long time nothing was done. At last a distinguished South African astronomer wrote to the City Council, pointed out the mistake, and suggested that the name should be changed to the correct Canopus.

I have actually seen the Council's reply. 'The name Canopsus has appeared on

*The Canopsus Street sign near
Pretoria as I saw it in 1975 —
so they did finally alter it.*

all our maps,' the letter runs, 'and cannot be altered now. We suggest that instead, you change the name of the star.'

I hardly think that this would have appealed to the astronomical fraternity. But they did make the alteration in the end.

52 ASTRONOMY AND THE LAW

On the whole, astronomers are friendly folk. They may—and do—disagree, but generally the arguments are civilised and free from personal animosity. Of course there are exceptions, but it is rare indeed to find any lawsuits brought in connection with astronomical matters. One or two have concerned the ownership of fallen meteorites, but to my mind the weirdest of all legal cases in astronomy was fought out in West Germany in 1953. The defendant was a lawyer named Godfried Büren, who was firmly under the impression that the Sun is cold rather than hot, and had offered 25,000 German marks (around £2,000 at that time) to anyone who could prove him wrong.

The idea of a cold Sun is not new. Even Sir William Herschel, that greatest of all observers, considered that below the bright surface there was a cool, temperate region which was very probably inhabited. Much later—in 1947—came a theory by another lawyer, this time from the Argentine. His name was Navarro, and in his view the Sun was not even solid, so that a well-aimed space-craft would be able to go straight through it. So far as I know, the only modern champion of the cold-Sun theory is a Sussex vicar, the Rev P.H. Francis, who holds a mathematics degree from the University of Cambridge. He has given various reasons for his support of this theory. For instance, he points out that when an electric fire is switched on, the bars become hot; but the power is supplied by an electrical generator miles away, and the generator itself is certainly not hot. Also, tea or coffee in a vacuum-flask will retain its heat, which cannot pass through the millimetre or two of vacuum between the inner and outer parts of the flask. There are 93 million miles of vacuum between ourselves and the Sun. . . . Like Sr Navarro, he too is doubtful whether the Sun is a real body. He believes that it may merely represent the gravitational centre of the Solar System.*

Now let us return to Herr Büren. He issued his challenge, and probably did not expect it to be taken up—but it was, by the West German Astronomical Society. The case was duly heard. Unfortunately there does not seem to be any transcript of it—it would make fascinating reading—but in the end the verdict went against Herr Büren, and he was ordered to pay over the 25,000 marks. There, sadly, the story seems to end, and I have never been able to find out whether the Astronomical Society received the money.

It was a curious episode by any standards, but I for one am glad that the world still has its quota of what I have called 'independent thinkers'. We would all be poorer without them.

*Mr Francis' booklet *The Temperate Sun* is well worth reading. I have described this and other unorthodox theories in my own book, *Can You Speak Venusian?* (Ian Henry Ltd, 1977).

53 PTOLEMY'S RED SIRIUS

The last of the great astronomers of Classical times was Claudius Ptolemæus, better known to us as Ptolemy. He wrote a major book which has come down to us by way of its Arab translation as the *Almagest*, and it is in the nature of a compendium; without it, we would know much less about ancient science than we actually do. There have been periodic attempts to discredit Ptolemy, and to claim that he was a copyist at best. Frankly, this seems highly improbable. In addition to being a good theorist (despite his belief that the Earth lay in the middle of the universe) he was a very competent observer. His star catalogue is of tremendous value. True, it was based on an earlier catalogue by the Greek astronomer Hipparchus, but Ptolemy extended it and improved it as well as making major contributions on his own account.

Generally, Ptolemy's descriptions of stars are reliable, but there is one real mystery. He described Sirius as red—and he was not alone. Earlier, Lucius Seneca had called Sirius 'deep red', and the same statement is to be found in ancient Egyptian and Assyrian texts. Today Sirius is pure white; if you doubt me, go and look. You will see Sirius well for several months each winter.

When low down, of course, it twinkles strongly. Twinkling has nothing directly to do with a star itself; it is due to the Earth's dirty, unsteady atmosphere, which splits up the starlight passing through. The lower the star, the greater the twinkling, because the light is reaching us through a thicker layer of air. Sirius is the supreme twinkler because it is so bright; it can also flash in various colours, but not when it is high up; and from Ptolemy's home in Alexandria, Sirius rises high in the sky. Twinkling, then, cannot be the answer.

It is most unlikely that Sirius itself has altered in hue. It simply isn't that kind

Impression of Ptolemy, although whether it is really like him we shall never know.

of star. But it has a dim companion, only 1/10,000 as bright, which is a very small, amazingly dense White Dwarf. A white dwarf is a bankrupt star which has used up all its nuclear energy, and has previously passed through the stage of being a red giant. There have been suggestions that in Ptolemy's time the companion was bright and red instead of faint and white. But this will not work either, because the time-scale is all wrong. A change of that sort takes tens of thousands of years at least, and we know that Sirius was white in the 10th century AD because the Arab astronomers said so. Moreover, a combination of the present Sirius and a brilliant red companion would add up to something bright enough to be seen in broad daylight, which again cannot have been the case.

On the whole, then, there seems no escaping the conclusion that on this occasion Ptolemy was either wrong or (more probably) misinterpreted. Yet why did other astronomers of ancient times agree in calling Sirius red? It is certainly a problem, but not one which we can ever hope to solve now.

54   INTRUDERS INTO THE ZODIAC

It is quite remarkable how even in 'this day and age', to use one of those infuriating modern expressions, people still confuse astronomy with astrology. I have never understood how this can happen. Of course, astrology—the superstition of the stars—was once regarded as a true science, but to my mind it proves only one scientific point: 'There's one born every minute'!

In astrology, the positions of the Sun, Moon and planets at the time of a person's birth have important effects upon his (or her) character and subsequent career. The belt round the sky marking the Zodiac is divided into 12 constellations, which are very different in size and prominence; thus Scorpius and Gemini are rich and brilliant, Cancer and Libra formless and obscure. In any case, the constellations are purely line-of-sight effects.

There is also the awkward fact (for astrologers) that the Zodiac also contains part of a thirteenth constellation, Ophiuchus (the Serpent-Bearer), which intrudes into the Zodiacal band between Scorpius and Sagittarius. Neither do the planets keep strictly to one circuit, and, for instance, Saturn lay in the constellation of Cetus (the Whale) for part of the year 1967. Pluto, whose orbit is inclined to ours at an angle of 17 degrees, can move well away from the actual Zodiac.

Astrologers, of course, ignore this sort of thing, presumably because their gullibility is equalled only by their ignorance of the fundamentals of astronomy. It is also significant that when Uranus, Neptune and Pluto were discovered in telescopic times (in 1781, 1846 and 1930 respectively) they apparently had no difficulty in fitting into the general astrological scheme. There is another problem today, because evidence is accumulating that Pluto is not worthy of being ranked as a true planet. Even when lumped together with its satellite, Charon, it is much less massive than the Moon.

All in all, astrology is harmless enough provided that it is treated as a parlour game, and confined to circus tents and the remaining seaside piers. It is also true that some astrological forecasts turn out to be correct. It would be surprising if they did not; as a famous judge once said, 'it is impossible always to be wrong' (though some modern politicians have made noble efforts to disprove this theory). But it must be said that there is absolutely no basis for astrology. It is a relic of the past, and it has no place in the modern world.

55   THE COAL SACK

Britons (and North Americans) are very familiar with the Great Bear or Plough; its Latin name is Ursa Major, and Americans often refer to it as the Dipper. Over London it never sets, and there are few people, even non-astronomers, who fail to recognise it.

In Australia, New Zealand or South Africa the same is true of Crux Australis, the Southern Cross. It never rises over Europe, which is a pity. Quite frankly it is not like an X (not nearly so X-formed, indeed, as our Cygnus, the Swan), and it is more like a kite, but it is the richest of all the constellations as well as being the smallest. Of its main stars, three are brilliant; Acrux or Alpha Crucis is a lovely double, while Beta and Gamma Crucis are also very prominent. Note that Gamma, unlike its companions, is strongly orange. The fourth member of the 'kite', Delta Crucis, is much fainter, and rather spoils the symmetry.

If you look just outside the main 'kite', near Acrux and Beta Crucis, you will make out what seems to be a barren region. Use binoculars, or a telescope, and you will see that there is an area which is virtually devoid of stars. Not surprisingly, it is known as the Coal Sack. There are other similar starless areas in other parts of the sky (in Cygnus, for example), and Sir William Herschel, discoverer of the planet Uranus, wondered whether they might be 'holes in the heavens', to use his own picturesque term. However, this is not so. They are dark nebulæ, and the Coal Sack is one of the most famous of them. It is rivalled only by the Horse's

The dark Horse's Head Nebula in Orion.

Head Nebula in Orion, but it is much closer to us; between 500 and 600 light-years. Its own diameter is probably about 65 light-years, though it is impossible to be sure.

A nebula is a cloud of dust and gas in space, and is a region in which fresh stars are being formed out of the interstellar material. The familiar nebulæ such as that in Orion's Sword are lit by stars in or near them; if the stars are hot enough, they excite the nebular material to a certain amount of self-luminosity. With the Coal Sack there are no suitable stars to make it shine, and it shows up as a dark mass, detectable only because it blots out the light of stars beyond.

This is why it appears starless, apart from a foreground star or two in front of it. But for all we know, there may be stars on the far side of the Coal Sack which light it up, so from another vantage point in the universe it could look bright. To us, however, it is—well, the Coal Sack; and it is easy to understand why even the great Sir William Herschel mistook dark nebulæ for holes in the heavens.

56 THE STRANGE FATE OF EDUARD VOGEL

One of the most extraordinary careers in the history of astronomy is that of Eduard Vogel. It led him to an early death—at least, so we assume; just what caused him to act in the way he did remains a complete mystery.

His early life was conventional enough. He was born in 1829 at Krefeld, son of the headmaster of the local school. He turned his attention to astronomy, and went first to Leipzig University where he studied under Heinrich D'Arrest (later to become famous as being Galle's assistant on the night of the discovery of Neptune). Vogel then went on to Berlin, where the Observatory director was Johann Encke. Vogel received his doctorate, and began serious work in computing the orbits of comets and minor planets. Then, in 1851, he went to England as an assistant at a private observatory in Regent's Park. (How times have changed! Anyone standing in Regent's Park today would find it impossible to see any stars at all.) From here, Vogel was the first to detect Encke's Comet at its return in 1852.

He seemed set for a successful career as an astronomer, but then came a diversion. In 1852 the British Government became alarmed at the fate of an expedition which had been sent to the then unexplored region of the Sahara Desert. One member of the expedition was a famous explorer, Heinrich Barth; it was believed that his companions had died, and that Barth was alone somewhere in the desert. Vogel was asked to go and look for him. It seemed strange to select a young astronomer for such a task, and even stranger for Vogel to accept; but he did, and in February 1853 he left for Africa. He never returned.

In October 1853 he wrote a letter from southern Libya, addressed to the German explorer Alexander von Humboldt, in which he described some observations made in the desert; these were actually printed in a leading German astronomical periodical in the following year. And certainly he found Barth; he met him in January 1855 on the shores of Lake Chad. By that time Vogel had joined with two young English soldiers, named Maguire and Church. The party had dinner together, and it is said that Barth even reproached Vogel for not having brought along some good German wine!

They stayed in the native village for a week, and then split up. Barth returned to England, accompanied by Church; later he published a book, *Travels and Discoveries in North and Central Africa*. Maguire went off on his own, and is believed to have been murdered by tribesmen in the Sahara. What, then, of Vogel? There seemed no reason why he should not return to his home and to astronomy; instead he continued his travels alone. The last glimpse of him shows him in the Upper Congo some time during September 1856. After that—silence.

It is hardly likely that he survived, and we can only speculate as to how he met his death. Astronomy was the poorer for his loss; had he returned with Barth he could so easily have made his mark in the scientific world.

57   AN OCEANIC MARS

One of the strangest theories about Mars was produced in 1901 by a German writer, Ludwig Kann of Heidelberg. He published it in a small book which is now very rare, but is certainly a true curiosity.

According to Kann, Mars is completely covered with water! After nightfall mists and clouds will form in the essentially Earth-like atmosphere, and there will be abundant rainfall, but the atmosphere itself is neither warmed nor cooled by radiation from any continents, so that storms and even winds are unknown. 'Naturally, then, the Martian ocean, far from being disturbed like the waters of the Earth, stretches in a mirror-like expanse, and masses of seaweed develop, covering the whole surface like a carpet. The regions of the Martian ocean which

Mars, from Mariner 7 (1969) at 535,650 miles. The bright ring at the upper right is Elysium.

are thus covered with uniform thick vegetation show a yellowish-red colour when seen from Earth; these are the so-called continents.

'The temperatures of the different regions of the Martian ocean will be regulated by the ocean currents, which are not diverted by any obstacles and are straight throughout their length; lifting the weeds in their path, they produce canals in the covering vegetation. At the ends of their courses, the currents broaden, and at the places from which the currents originate the ocean is disturbed over a large area, so that the vegetation cover is broken up or driven away to a degree depending upon the intensity of the movement; the dark patches which result are the so-called seas.'

Kann goes on to give equally remarkable explanations of the apparent twinning of the canals as reported by Schiaparelli ('counter-currents') and the seasonal variations observed ('the greater or lesser density of the vegetation cover of the Martian ocean'). He compares Mars today with the state of the Earth in the Carboniferous period, when the coal deposits were being laid down, and concludes: 'We may assume the existence of organisms similar to those which were found on Earth at that epoch. It is, then, by studying the organisms which have been so admirably preserved in the deposits of those remote times that we can form an idea of the flora and fauna now living on Mars; we can see them and analyze them. In the Martian ocean we see nearly all the principal organisms of the terrestrial seas of today, apart from the reptiles and mammals; we see the floating lands covered with thick forests made up chiefly of calamites, magnificent tree-like ferns and a great variety of forms whose strong and widespread roots help in consolidating the mass. At the edges of the freshwater lakes, in the middle of the forests extending over a floating land above the ocean, we see signs of animal life such as scorpions and spiders running along the ground, while superbly-coloured dragonflies of enormous size fly in the air. The lake-floor is covered with small mussels; there are fishes living on the larvæ of insects, together with crabs of various kinds. The whole world is bathed in a dazzling light, radiating from a cloudless sky. Not the slightest whisper; and absolute silence reigns everywhere. Man has not appeared; neither have mammals; no birds have ever been created on Mars. But sooner or later a catastrophe will happen, and this happy land, with all its living things, will meet its doom at the bottom of the ocean. Then, after a long pause, Man will appear on Mars, to revive the lands and use water-power to warm his houses, and make his fabrics and his machines.'

If this had been correct, then the Viking soft-landers of the 1970s would have had to equip themselves with floats. But I fear that the Mars we know is very different from the idyllic world pictured by Herr Kann.

58   THE SOUTH POLE OF THE SKY

One of the most famous stars in the sky is Polaris or the Pole Star, in the constellation of Ursa Minor (the Little Bear). It is 6,000 times as powerful as the Sun, but it is 680 light-years away, and so it shines as a star of only the second magnitude—conspicuous enough, but by no means outstanding. To us, its importance lies in the fact that it is within one degree of the north celestial pole, so that it seems to remain almost stationary, with the entire sky revolving round it once in 24 hours.

The Earth's axis of rotation passes through the north and south poles, and also through the centre of the globe. We may imagine that the Earth is surrounded by a transparent sphere, which we call the celestial sphere. Naturally it does not really exist (though the ancients thought it did), and is merely a convenient fiction. The celestial poles are the points where the Earth's axis, if prolonged, would hit the celestial sphere.

Polaris is useful as a 'skymark'. Its apparent altitude above the horizon is equal to the observer's latitude on the Earth; thus from latitude N.51° the altitude of Polaris is also 51° (for the moment we can ignore the fact that Polaris is not exactly on the polar point). Viewed from the North Pole, Polaris would have an altitude of 90°; that is to say, it would be directly overhead. From the equator (latitude 0°) Polaris has an altitude of 0°, so that it is exactly on the horizon; and from latitudes south of the equator Polaris cannot be seen at all.

Polaris has always been of the greatest value to sea navigators; but what happens when we go south of the equator? Polaris is out of view, and we must look instead toward the south celestial pole. Unfortunately this lies in the obscure constellation of Octans, the Octant, which contains no bright object, and the south polar star, Sigma Octantis, is none too easy to see with the naked eye, so that even moderate moonlight will drown it. It is much less powerful than Polaris, and is only seven times more luminous than the Sun; its distance is 120 light-years.

The whole south polar area is very barren. The best way to locate it is by using the brilliant star Achernar, in Eridanus (the River) and the Southern Cross, both

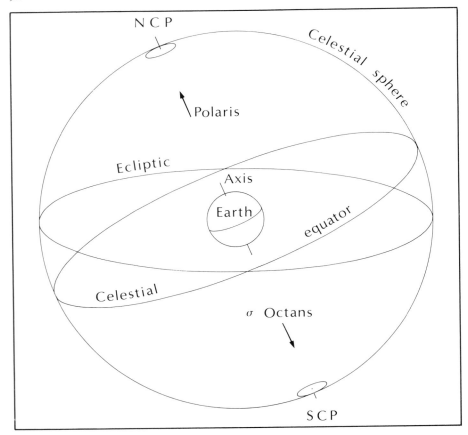

The celestial sphere showing north and south celestial poles.

of which are familiar to any star-gazers who live in countries such as Australia and New Zealand. The south celestial pole lies almost directly between them. The nearest reasonably bright star to the pole is the third-magnitude Beta Hydri, but it is not prominent, and in any case it is almost 14 degrees from the polar point.

Because of the phenomenon known as precession, the celestial poles shift slowly from year to year, and eventually the southern polar star will be the brilliant Canopus. But this will not happen for thousands of years yet, so I am afraid that in our time we must be content with the admittedly inadequate Sigma Octantis.

59 JOHN HERSCHEL AND THE SOLITARY STAR

During the 1830s Sir John Herschel, son of Sir William, spent some years at the Cape of Good Hope. He took with him a powerful telescope (now to be seen at the Old Royal Observatory in Greenwich) and carried out a detailed survey of the southern stars, which can never be seen from Europe. In 1838 he set out for home. The voyage took a long time, and some naked-eye observations which he made seem worthy of being put on record.

Herschel looked at the star Alphard, or Alpha Hydræ. It is reasonably bright, and the modern estimate of its magnitude is 1.98, almost exactly equal to the Pole Star. It is orange, and 700 times as luminous as the Sun; its distance from us is 85 light-years. It is often called the 'Solitary One', because it lies in a barren part of the sky. Actually it is only nine degrees south of the equator, so that it is visible from almost all inhabited parts of the world; you can find it easily enough by using the Twins, Castor and Pollux, as direction-finders.

Alphard is not known to be variable in light, but Herschel believed that it did change, and he made estimates of its brightness by comparing it with other stars: Beta Aurigæ or Menkarlina (magnitude 1.90), Wezea or Delta Canis Majoris (1.86), Polaris (1.99), Algieba or Gamma Leonis (also 1.99), Koo She or Delta Velorum (1.96) and Castor (1.58). Herschel made his first estimate on March 21 and his last on May 12 1838. His ship docked on May 15, and apparently he made no further observations.

The results were rather interesting. His diary reads as follows:

21 March Inferior to Delta Canis Majoris, brighter than Delta Velorum and Gamma Leonis.

7 May Fainter than Beta Aurigæ, very obviously fainter than Gamma Leonis or Polaris; 'now a decidedly insignificant star. . . . Moon, and low altitude, but it leaves no doubt in my mind of the minimum being nearly attained.'

8 May 'Very decidedly inferior to Gamma Leonis. . . . I incline to place the minimum as last night as it is tonight brighter than Beta Aurigæ at certain intervals of its twinkling. . . . On the whole after many comparisons it is rather inferior to Beta Aurigæ.'

9 May Inferior to Gamma Leonis, but 'evidently on the rapid increase. The

Sir John Herschel.

minimum is certainly fairly passed and the star is rapidly regaining its light.'

10 May 'Much inferior to Gamma Leonis, rather inferior to Beta Aurigæ. It is still about its minimum.'

11 May 'Brighter than Beta Aurigæ no doubt; Beta much higher.'

12 May 'Castor and Alpha Hydræ nearly equal.'

Herschel was an experienced observer, but since then there have been no suggestions that Alphard is markedly variable. It is awkward to estimate because of the lack of nearby comparison stars; it may fluctuate slightly, but certainly by no more than a few tenths of a magnitude at most.

60 ASTRONOMERS ARE HUMAN!

Astronomers are usually regarded as being the reverse of accident-prone, and able to avoid elementary mistakes. Generally speaking this is true, but there have been some curious episodes now and then. Perhaps the best was the report made by an earnest observer in the 1790s. Using a reflecting telescope, he reported that he had seen large creatures walking about on the surface of the Moon. Unfortunately, they turned out to be ants in his eyepiece.

In more modern times there have been some remarkable errors of judgement. Pre-eminent, probably, is the 600-ft steerable radio 'dish' telescope which was to be built at Sugar Bowl in the United States. When construction had been in progress for some time, and had cost a great many dollars, it was discovered that the structure could not possibly work, and it was quietly abandoned. Those responsible would vastly prefer to forget all about it, so I will say no more here.

But what about the connection between 4 Herculis, a hot, bluish-white star, and the ordinary, common-or-garden match? It sounds improbable, but the story of the 'Match Parade' certainly merits a place in the history of astronomy.

It began in the 1960s with some work carried out at the French observatory at Haute Provence. This is one of the major observatories in Europe, and is equipped with a fine 76-in reflector. Astronomers there were busily engaged in studying the spectra of some special kinds of stars when they made, or thought they made, a very remarkable discovery.

Normally, the spectrum of a star shows a rainbow background crossed by dark absorption lines, each of which is due to some particular element or group of elements. In this way we can tell what substances are present there (despite August Comte). In studying the spectrum of a star with the catalogue number of HD 117043, a dwarf star of spectral type G6 and therefore rather smaller and cooler than our Sun, the French investigators suddenly detected lines due to the element potassium which were not dark, but bright. Now, bright potassium lines are the last things to be expected with a yellow dwarf star, and the astronomers were badly puzzled. A couple of years later, another 'potassium flare' was detected in the spectrum of the orange star D 88230. By now the French observers were really intrigued, and began a systematic search, concentrating upon dwarf stars and also stars with unusual spectra. One more potassium flare was reported, this time in the spectrum of the bluish-white star 4 Herculis, and the mystery grew.

What could be the answer? Potassium flares in such stars seemed to make no sense at all; yet there they were. Apart from the strong, broad potassium lines, the spectra of the stars appeared to be quite normal. It was all most peculiar.

The search shifted to America. At the Lick Observatory, F. Wing, N. Peimbert and H. Spinrad used the 36-in reflector in a systematic hunt, covering 162 bright stars. Absolutely nothing abnormal was found; potassium flares were conspicuous only by their absence. Finally, the Americans became decidedly suspicious, and made some unorthodox tests.

Using the Lick equipment, they took spectra of book, kitchen and safety matches. Now, matches contain potassium. The spectra showed broad, bright lines—exactly the same as those which the French observers had reported in the spectra of the stars. Similar tests were then carried out at Haute Provence. There seemed to be no difference between American and French matches; potassium lines showed up each time. Finally, the Americans found the answer.

Suppose an observer, having finished his exposure at the telescope, strikes a match to light a cigarette? Probably there was no thought that the stray light could reach the equipment, but it was possible for off-axis light to enter by reflection from a glass plate used as a guiding device at the observing platform. And this gave the solution. What the French had recorded was not potassium in the stars, but potassium in the observer's matches. Further studies left no room for doubt.

It is only too easy to laugh, but it does show that even skilled professional astronomers are not infallible. Even Homer nods occasionally.

61   LIFE ON A COMET

The first British populariser of astronomy was James Ferguson, who lived from 1710 to 1776. He was self-taught, but wrote very well indeed, and his *Astronomy Explained on Sir Isaac Newton's Principles* ran to many editions. Some of it makes amusing reading now. For instance, listen to what Ferguson has to say about comets:

'The comets are solid opaque bodies with long transparent trains or trails, issuing from the side which is turned away from the sun. They move about the sun in very eccentric ellipses, and are of a much greater density than the earth; for some of them are heated in every period to such a degree, as would vitrify or dissipate any substance known to us. . . .

'It is believed that there are at least 21 comets belonging to our system, moving in all sorts of directions; and all those which have been observed, have moved through the ethereal regions, and the orbits of the planets, without suffering the least sensible resistance in their motions; which plainly proves that the planets do not move in solid orbs. . . .

'The extreme heat, the dense atmosphere, the gross vapours, the chaotic state of the comets, seem at first sight to indicate them altogether unfit for the purposes of animal life, and a most miserable habitation for rational beings; and therefore some are of opinion that they are so many hells for tormenting the damned with perpetual vicissitudes of heat and cold.* But when we consider, on the other hand, the infinite power and goodness of the Deity; the latter including, the former enabling, him to make creatures suited to all states and circumstances; that matter exists only for the sake of intelligent beings; and that wherever we find it, we always find it pregnant with life, or necessarily subservient thereto. . . it seems

Donati's Comet of 1858, from an old woodcut.

highly probable, that such numerous and large masses of durable matter as the comets are, however unlike they be to our earth, are not destitute of beings capable of contemplating with wonder, and acknowledging with gratitude, the wisdom, symmetry, and beauty of the creation; which is more plainly to be observed in their extensive tour through the heavens, than in our more confined circuit. If farther conjecture is permitted, may we not suppose them instrumental in recruiting the expended fuel of the sun, and supplying the exhausted moisture of the planets? However difficult it may be, circumstanced as we are, to find out their particular destination, this is an undoubted truth, that wherever the Deity exerts his power, there he also manifests his wisdom and goodness.'

Much later, in 1877, appeared the English translation of a book by the French astronomer Amedée Guillemin, *The World of Comets*. He devotes a whole chapter to the question of the possible habitability of comets, and though he finally decides that such a thing is most unlikely he spends five pages in discussing it. We have come a long way since then.

*Ferguson is here referring to the Rev William Whiston, a contemporary of Newton, who was responsible for an end-of-the-world scare in 1736—I have described it fully in my book *Countdown* (1983).

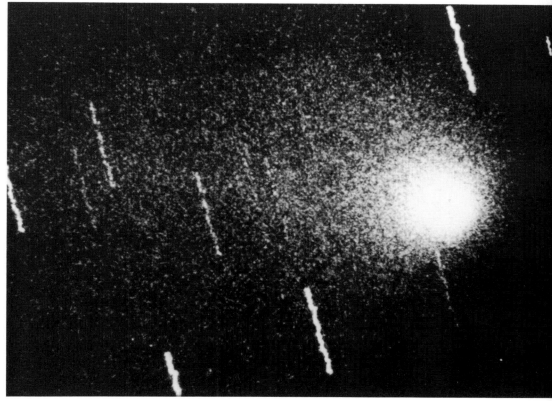

Comet IRAS-Araki-Alcock of 1983. This was a very small comet, but came relatively close to Earth and was visible to the naked eye for a few nights.

62   NEWS FROM VEGA

Vega, or Alpha Lyræ, is the fifth brightest star in the sky. From England it is almost overhead during summer evenings, and is easily recognised, not only because of its brilliance but because of its strong blue colour—indicating that its surface is very hot. It is 50 times as luminous as the Sun, and has a diameter of rather less than two million miles. Its distance from us is 26 light-years, which works out to around 150,000 million miles.

Very recently (October 1983) a remarkable discovery was made. It was due to IRAS, the Infra-Red Astronomical Satellite, which was launched on the preceding January 26. Its mission was to collect long-wavelength radiations from space, using a special infra-red telescope. It detected some 200,000 sources, some of which are new, and it has also been responsible for the discovery of five comets. But it was not designed to study individual stars in our own part of the Galaxy, and when two of the IRAS scientists, Drs Hartmut Aumann and Fred Gillett, turned the equipment toward Vega they did so merely to check the adjustment of the infra-red telescope.

To their surprise, they found that they were detecting a great deal of infra-red radiation from the neighbourhood of Vega. This could only mean that Vega was associated with a swarm of solid particles. Further studies showed that the radiation is coming from an extended region around Vega, stretching out to some 80 astronomical units from the star—that is to say, 80 times the distance between the Earth and the Sun, which amounts to almost 7,500 million miles. IRAS was also able to measure the temperature of the material: it was about – 300 degrees Fahrenheit.

From theoretical considerations, Aumann and Gillett found that the material is not simply 'cosmic dust', of which there is plenty scattered around the Galaxy. The material around Vega may be a planetary system in the process of formation. There may even be bodies comparable in size with the planets of our own Solar System, and it seems likely that the total mass of material is about equal to that of all the Sun's planets combined.

Obviously this discovery has far-reaching implications. Vega is an ordinary star, by no means exceptional in size or luminosity. It is younger than the Sun; the age of the Sun is about 5,000 million years, that of Vega no more than 1,000 million years or so, in which case it is hardly likely that any of Vega's planets (if they really exist) are old enough for advanced life to have developed there. But one never knows; and if Vega has a system of this kind, there is every reason to believe that the same is true of many other stars.

So far this is as much as we know. Further studies, either from IRAS-type satellites or from telescopes based on Earth, may help us to decide how large the orbiting bodies are, and whether the material is in the form of a sphere or a ring. But a start has been made, and astronomers will be keeping a very close watch upon this brilliant blue star.

63   METEORS OVER THE MOON?

With its low escape velocity of only 1½ miles per second, the Moon could not possibly hold on to anything much in the way of an atmosphere. This has been known for a very long time. Yet until relatively recently it was still thought that there might be a trace of atmosphere remaining. For instance, in 1892 the American astronomer W.H. Pickering observed an occultation of Jupiter by the Moon, and recorded a dark band crossing the planet's disk, tilted with respect to the Jovian belts; this he attributed to a lunar atmosphere with a ground density of about 1/1800 that of the Earth's air at sea level. He also considered that this atmosphere was frozen solid during the lunar night.

Few people agreed with him, and it was clear that the density of any atmosphere must be far less than this. In 1949 the Russian astronomer Y.N. Lipski announced the definite detection of an atmosphere about 1/10,000 that of the Earth. This is, of course, what we normally call a good laboratory vacuum; but it would have one important result—it would mean that there might be meteors observable from the Moon.

In 1952 I discussed this problem with Dr Ernst J. Öpik, a well-known Estonian astronomer who had been working at the Armagh Observatory in Northern Ireland. In a letter to me, Öpik wrote as follows:

'Lunar meteors are quite probable. Considering the surface gravity of the Moon, which leads to a six times' slower decrease of atmospheric density with height, the length of path and duration of a meteor trail on the Moon will be six times that on Earth, if a thin atmosphere exists. However, meteors the size of fireballs will penetrate the lunar atmosphere and hit the ground. The average duration of a meteor trail on the Moon will be two to three seconds (as against half a second on Earth), and each trail should end with a flash when the meteor strikes the ground (because all meteors which can be observed in the lunar atmosphere from such a distance must be large fireballs). The average length of trail would be 75 miles, or one minute of arc—1/30 of the Moon's diameter—and the meteors would therefore be very slow, short objects.'

This sounded convincing enough. In America, members of the Association of Lunar and Planetary Observers made a systematic search, and reported quite a number of objects which were regarded as lunar meteors. The average path-length was indeed 75 miles; some left brief trails, and all were rather faint. But. . . we now know that to all intents and purposes the Moon has no atmosphere at all. Lunar meteors are absolutely out of the question. I fear that we have yet another case of innocent self-deception. Whatever the American observers saw, they were most certainly not meteors in the non-existent atmosphere of the Moon.

64   SERPENT IN THE SKY

Comets are the most erratic members of the Solar System, and of bright comets only Halley's is a regular visitor to the neighbourhood of the Sun. Theories about comets in general abound. A new one has now been proposed by two highly respected astronomers at the Royal Observatory, Edinburgh: Drs Victor Clube and Bill Napier. If they are right, then almost everyone else has been wrong, which is, of course, perfectly possible!

Conventionally, it is assumed that comets come from the 'Oort Cloud', a reservoir of comets moving around the Sun at a distance of about a light-year (it is named because it was first proposed by the Dutch astronomer Jan Oort). When a comet is perturbed for any reason, it falls inward towards the Sun, and it may be so violently perturbed by a planet—usually Jupiter—that it becomes a short-period comet. Each time the comet then passes perihelion, some of the ices in its nucleus evaporate, which means that it must be short-lived on the cosmical scale, and it will finally lose all its gas, leaving nothing but a small, dead core.

It has also been assumed that comets are genuine members of the Solar System, and that the Oort Cloud dates back several thousands of millions of years. Clube and Napier disagree. They believe that the original Oort Cloud must long since have been dissipated, and that a new Cloud is picked up by the Sun when it passes through one of the spiral arms of the Galaxy. In fact, comets are, basically, interstellar wanderers. When the Sun collects a new swarm of comets during its passage through a spiral arm, it follows that for millions of years afterwards the numbers of comets in the Solar System are unusually great.

What about dead comets? Clube and Napier believe that these are nothing more nor less than Apollo-type asteroids—small bodies, a few miles across, which have eccentric orbits bringing them close to the Earth. One such asteroid, Hephaistos, discovered in 1979, is around six miles in diameter, and moves in an orbit similar to that of Encke's Comet, which has a period of 3.3 years. The orbit is also much the same as that of the meteors of the annual Beta Taurid shower, and some time ago F.L. Whipple suggested that all three resulted from the break-up of a much larger body in the third millenium BC.

Napier and Clube then turn to the near-present, and look at mythology and history. Men have always been afraid of comets (the fear is not dead even yet in some countries). It is therefore suggested that in near-historical times a very large comet, which they call the Cosmic Serpent, was forced into an Apollo-type orbit. It made regular close approaches to the Earth, and would then have been brighter than the full moon, with a magnificent tail stretching right across the sky. It would be accompanied by superb meteor displays, and, more importantly, by debris hitting the Earth. This would mean that whenever the comet appeared, disasters would be likely. There would also have been splendid displays of Zodiacal Light, which is due to sunlight reflected from interplanetary material, and it is even possible that the denser part of the meteor stream would dim the Sun over a period of several consecutive weeks.

However, this situation did not last. Gradually the Serpent lost its gas; it faded from view, the meteor displays declined, the Zodiacal Light became a feeble glimmer, and the period of disastrous impacts was over, at least for the moment. If so, then where is the Cosmic Serpent now? Clube and Napier believe that it is Hephaistos, nothing more than the dark remnant of its former self.

Well, that is the theory. It is revolutionary and unorthodox; it is speculative; but it is worth taking very seriously, and it is an honest effort to link astronomy with palæontology, archæology, geology, mythology and even history. If it is correct, then at the moment we are in a 'safe' period, and the chances of an Apollo strike are low. But eventually the Sun will again traverse one of the Galaxy's spiral arms, the Oort Cloud will be replenished, and the results will be spectacular. In short, the theory means that what happens on Earth is controlled largely by what happens in the outer Galaxy.

At least there are no Cosmic Serpents today, which is something for which mankind should be duly thankful.

65 THE COSMIC ZEBRA

Upon first seeing a zebra, one is inclined to ask: 'Is it a white animal with black stripes, or a black animal with white stripes?' Zoologists have no difficulty in providing an answer, but astronomers have been faced with roughly the same problem, and have solved it only very recently. Our cosmic zebra is the outermost of the large satellites of Saturn, Iapetus.

It was discovered in 1671 by G.D. Cassini, so that it was the second member of Saturn's family to be found (Titan was the first). Iapetus moves round Saturn at a mean distance of 2,200,000 miles, in a period of 79 days 7 hours 56 minutes. Seen from Saturn it would have an apparent diameter of less than two minutes of arc; its real diameter is approximately 900 miles, so that it is considerably smaller than our Moon.

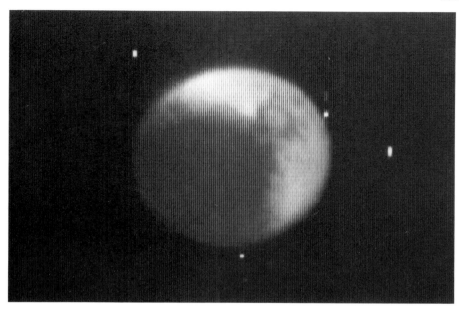

Iapetus, showing its bright and dark areas. The sunlight is coming from the lower left.

After Cassini had discovered Iapetus, he followed it carefully, and noted something very odd. When west of the planet, Iapetus was easy to see, with a magnitude of about 10; but when east of the planet, Cassini could not see it at all. In fact the magnitude drops considerably, though any modern telescope with an aperture of 6 in will follow it comfortably enough throughout its orbit.

The variations were regular, and it was assumed (correctly) that the rotation period is the same as the revolution period, so that Iapetus always keeps the same face turned toward Saturn. In this case, either the shape must be irregular, or else the two hemispheres must be of unequal reflecting power—one bright and one very dark indeed. Hence the zebra comparison. Was Iapetus white, with a dark strain, or dark, with a white coating?

The clue was given by the determination of the mean density of the globe. Iapetus is only about 1.2 times as dense as water. Therefore, it must contain a good deal of ice mixed with its rock. Were it made up principally of rock, it would be much denser and more massive than it actually is. Therefore it is the dark hemisphere which is the 'stain'.

Unfortunately Iapetus was not as well mapped by the Voyager probes as some of the other satellites, but craters were found on both the bright and the dark areas. It was also seen that the boundary between the two regions is not sharp, and there is a transition zone about 150 miles broad. It is the 'leading' side which is dark, and this has led to the suggestion that the stain could be due to dusty material coming from the outermost satellite, Phœbe, which is very small and apparently darkish. But this seems unlikely; Phœbe is a long way out, and probably could not produce enough 'dust'.

On the whole, it seems that the dark material has welled out from inside

Iapetus. It could be a mixture of ammonia, soft ice and dark material of some kind. We know nothing about its thickness; it may go down for only a few inches, or it may be deep. Carl Sagan has even suggested that it could be organic, though this seems frankly improbable.

So Iapetus remains a puzzle, and not until another space-craft passes by it are we likely to find out just why it is so unlike any other satellite in the Solar System.

66   W.H. STEAVENSON: DOCTOR AND ASTRONOMER

In 1934, when I was 11 years old, I had managed to save up the princely sum of £7 10s, which in those days was a great deal. I was determined to buy a telescope, and equally determined that it should be a good one. Fortunately I was lucky. A relative of mine had a passing acquaintance with Dr William Henry Steavenson, a GP in Outer London, whose fame as an amateur astronomer was world-wide. Dr Steavenson came down to my home in East Grinstead, and chose my telescope for me: a fine 3-in Broadhurst Clarkson refractor, which I still have and which I still use.

Steavenson was a remarkable man. A boyhood accident robbed him of one eye, but nevertheless he became a superbly accurate observer. He had no wish to become a professional astronomer; he took his medical qualifications, and set up in practice, but although he was an excellent doctor it is for his astronomical work that he will always be remembered. He was particularly interested in old novæ, which are the 'remains' of the short-lived brilliant stars which flare up now and again. He erected a modest observatory in his garden, but when he acquired a fine 30-in reflector he set it up at Cambridge University, where the authorities were only too glad to accommodate him. Later I often used to go there and observe with him.

Amateur though he was, Steavenson was often called in to help in professional programmes, particularly site-testing for new large telescopes. He was a leading authority on the equipment made and used by William Herschel, and indeed his knowledge of all the world's major telescopes was encyclopædic. He had a long association with the Royal Astronomical Society, and served as President in 1957-59; few 20th-century amateurs have done as much.

He was also closely connected with the British Astronomical Association, and was President in 1926-28; at various times he directed the Mars, Saturn and Observing Methods Sections. I believe that he went for more than 30 years without missing a single council meeting. It seemed strange indeed when failing health forced him to give up; he retired to Swindon, and died there at the age of 81 in September 1975.

One of Steavenson's greatest merits was the encouragement he gave to young people—as I have excellent reason to know. He had a strong sense of humour, and to the end of his life he delighted in reading magazines intended for teenage boys! I remember, too, how he used to refer to his patients as 'beetles' when they came to see him when the sky was dark and clear. . . .Not that he ever neglected his practice—far from it.

There are few great amateur astronomers who have done as much for the science as W.H. Steavenson. He wrote one small book, *Suns and Worlds*, and many technical papers, but he will be remembered above all for his warm, genial personality.

67   URANIA'S MIRROR

Some time ago I had a letter from the Director of the Almonry Museum in Evesham, enclosing two coloured cards depicting constellations and asking me if I could identify them. Frankly I could not, but shortly afterwards I had a piece of luck. I was lecturing at Wansfell Residential College in Theydon Bois (as I have done every year since 1947!) and a member of the course produced a dozen similar cards. These gave me a clue, and subsequently I found an article about them in the well-known American periodical *Sky and Telescope* which completed the story.

The cards are known collectively as Urania's Mirror. The cards are opaque, but when backlit the stars show up as bright against the dark background, and the effect is beautiful. I have yet to see a complete set, though there is one in the Smithsonian Institution in America.

One intriguing aspect of the cards is that they show some constellations which are not now recognised. In fact, the whole sky was 'confused' for many years because astronomers had a habit of introducing new constellations, forming them

One of the pictures from Urania's Mirror, showing some of the now-rejected constellations such as Machina Electrica!

out of the older ones; this persisted until 1932, when the International Astronomical Union, the controlling body of world astronomy, lost patience and reduced the accepted number of constellations to 88.

In Urania's Mirror, therefore, we find some unfamiliar names: typical are Ballon Aerostatique, the Air Balloon (now included in Piscis Australis); Noctua, the Night Owl (borders of Hydra, Libra and Virgo); Sceptrum Brandenburgicum, the Sceptre of Brandenburg (Eridanus); Telescopium Herschelii, Herschel's Telescope (Auriga); Felis, the Cat (Hydra); and Psalterium Georgii, George's Harp (again Eridanus). Of these old groups, only one, Quadrans (the Quadrant) is remembered today, because the annual January meteor shower known as the Quadrantids radiates from the position where Quadrans was placed; it is now part of Boötes, the Herdsman.

Who produced the cards? All we know is that they were due to 'a lady', and there seem no obvious candidates. The only Englishwoman then known in the astronomical world was Mary Somerville, but she was a mathematician, and Urania's Mirror does not seem to fit in with her activities. So the mystery remains, but certainly Urania's Mirror is worth examining—if you are ever fortunate enough to come across it.

68 FOMALHAUT: A SECOND PLANETARY SYSTEM?

Fomalhaut, in the constellation of Piscis Australis (the Southern Fish) is the southernmost of the first-magnitude stars to be visible from Britain. It lies almost 30 degrees south of the celestial equator; from the London area it rises to a respectable altitude above the horizon, and may be found by using the stars of the Square of Pegasus as a guide, but from North Scotland it barely rises at all. Therefore, Britons do not appreciate its true brilliance. When seen overhead, as from Australia or South Africa, it is very conspicuous indeed; it is in fact the 18th brightest star in the whole sky, with an apparent magnitude of 1.16, about the same as that of Deneb in Cygnus. There is little to be said about the rest of the Southern Fish, which contains no other star as bright as the fourth magnitude.

At its distance of 22 light-years, Fomalhaut is a comparatively close neighbour; of the first-magnitude stars only Alpha Centauri, Sirius, Procyon and Altair are nearer. Fomalhaut is a white star of spectral type A, and it has 13 times the luminosity of the Sun.

Fomalhaut is notable as being the second bright star to be associated with detectable 'planetary' material. The first was Vega, where the material was discovered by the instruments on IRAS, the Infra-Red Astronomical Satellite. Following this unexpected development, other stars were examined. Some, such as Altair, were ordinary enough; but with Fomalhaut the same 'infra-red excess' as had been found with Vega was much in evidence. Again it is impossible to decide on the sizes of the pieces of material; but some of them may be of planetary dimensions, and Fomalhaut, like Vega, may be in the process of developing a system not very unlike that of the Sun.

It is not likely that life has had time to evolve there, because although Fomalhaut is not nearly so luminous as Vega it is still much more powerful than the Sun, and is presumably much younger. All the same, the chances of an Earth-type planet are greater than with Vega.

Yet the real significance of the discovery is that Vega is not unique. Fomalhaut has the same sort of surroundings, and if two of the closer bright stars are planetary centres the same is presumably true of many others. Whether we will be able to detect any of these planets visually seems doubtful as yet, but Fomalhaut, a mere 22 light-years away, may present us with a real chance in the not-so-distant future.

69   VENUSGLOW

I have always been intrigued by the reputation of Franz von Paula Gruithuisen, who was born in Castle Haltenberg in Bavaria in 1774 and died in 1852. He was a fairly good observer, who concentrated upon the Moon and planets, but his imagination was—to put it mildly—vivid. For instance, he discovered what he believed to be an artificial structure on the lunar surface, and described it as 'a collection of dark gigantic ramparts. . . a work of art'. Alas, nothing is there apart from a few low, haphazard ridges, and there is no question of any change, because a far better observer, Johann Hieronymus Schröter, had described the area correctly some time earlier. But it was with regard to Venus that Gruithuisen really excelled himself. In particular, he proposed a theory of the Ashen Light which caused his more sober-minded contemporaries to raise their eyebrows.

The Moon's unlit hemisphere is often made visible by light reflected from the Earth ('the Old Moon in the New Moon's arms'; see article 40 above). Telescopically, the same appearance can be seen for Venus, and is known as the Ashen Light. It is not so easy to explain; Venus has no moon, and the Earth could not illuminate it perceptibly, so what is the answer? Gruithuisen thought that he had found it. He noted that the Light had been observed in 1759 and in 1806, an interval of 47 terrestrial or 76 Cytheean years, and wrote:

'We assume that some Alexander or Napoleon then attained universal power. If we estimate that the ordinary life of an inhabitant of the planet lasts for 130 Venus years, which amounts to 80 Earth years, the reign of an Emperor of Venus might well last for 76 Venus years. The observed appearance is evidently the result of general festival illumination in honour of the ascension of a new emperor to the throne of the planet.'

Later, Gruithuisen had second thoughts, and wondered whether the Light might be due to the burning of large stretches of jungle to produce new farmland, adding that 'large migrations of people could be prevented, so that possible wars would be avoided by abolishing the reason for them. Thus the race would be kept united.'

There are, perhaps, certain objections to these ideas, but the Ashen Light has always been controversial. I do not believe that it is a mere contrast effect; I have seen it too often and too clearly, and the best theory is that it is due to electrical phenomena in the upper atmosphere of the planet. However, we must not dismiss Gruithuisen as a pure eccentric. He did make some useful observations, and he was the first to propose that the craters of the Moon were produced by meteoritic impact rather than vulcanism—a theory now very popular, particularly in the United States, though personally I beg to differ.

70   THE SOOTY STAR

One of the most remarkable stars in the sky is R Coronæ Borealis. It is never brighter than about the sixth magnitude, and is therefore on the fringe of naked-eye visibility, but it is of special interest.

To find it, first locate Corona Borealis, the Northern Crown. This is easy enough. Follow round the tail of the Great Bear, and identify the brilliant orange Arcturus in Boötes, the Herdsman. Corona lies not far away, and is made up of a semicircle of stars, of which the brightest, Alphekka or Alpha Coronæ, is of the second magnitude—about equal to the Pole Star. Inside the bowl of the crown binoculars will usually show two stars. One is of magnitude 6.5, and the other is R Coronæ. However, you will sometimes find that R Coronæ is missing. At unpredictable intervals it drops in brightness, taking only a few days or a week or two to fall to magnitude 15—well beyond the range of binoculars or small telescopes. After remaining faint for a while, it slowly and jerkily recovers its lost light, remaining more or less steady until the onset of the next minimum.

Why does R Coronæ behave in this peculiar way? Spectroscopic work has shown that it contains much less hydrogen than a normal star, but there is an excess of carbon. Apparently there are times when the carbon accumulates in the star's atmosphere as nothing more nor less than soot! In fact, the star retires behind a sooty veil, and much of its light is cut off. Then, when the soot disperses, the veil clears away and the star shines out once more.

R Coronæ-type stars are very rare. In fact, there are only five others which can become as bright as the ninth magnitude: RY Sagittarii, S Apodis, UW Centauri, RS Telescopii and SU Tauri. All are luminous and remote, and there seems to be a certain doubt as to where they fit into the general scheme of stellar evolution. Because their falls to minimum cannot be predicted, amateur astronomers do valuable work by keeping a watch on them, so that professionals, with their spectroscopic equipment, can begin their studies as soon as the fade starts.

So when the sky is clear and Corona Borealis is above the horizon, it is always worth using binoculars to take a quick look inside the bowl of the Crown. If you can see only one star instead of two, you will know that R Coronæ has started to put on a performance. But never fear—it will soon be back, and there is no danger that we will permanently lose sight of our sooty star.

71   A TUNNEL ON THE MOON?

On the dry plain of the lunar Mare Fœcunditatis, or Sea of Fertility, we find an interesting pair of small craters. Neither member of the pair has a diameter as great as ten miles. The eastern member is named Messier in honour of the great 18th-century comet-hunter. The companion was once named Pickering, but for some unexplained reason the name has been deleted from modern maps, and the western member is known as Messier A. They are notable because of a strange double ray extending westward from them in the direction of the eight-mile crater Lubbock. No rays are seen in other directions, so that either the ray material was ejected only over a narrow angle or else the surrounding surface was not solid enough to retain the ray material.

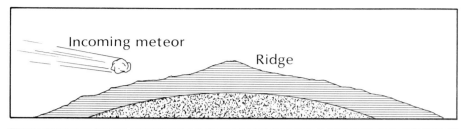

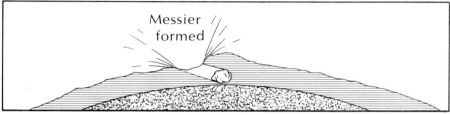

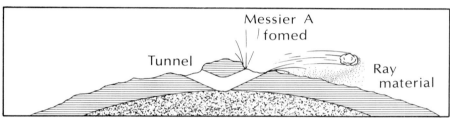

The lunar tunnel theory.

The first really great lunar mappers, W. Beer and Johann Mädler, wrote in 1837 that the two craters were exactly alike. 'Diameter, shape, height and depth, colour of interior are the same, and even the positions of the peaks; everything points to the fact that we have here either a most remarkable coincidence, or that some as yet unknown law of nature had been at work.' Today the two are not alike; A is triangular and Messier itself elliptical, while A generally looks the larger of the two. I say 'generally' because the two appear to change in size and shape according to the angle at which the sunlight strikes them. There is no suggestion of real change, either short-term or since the description given by Beer and Mädler—who, magnificent observers as they were, used a modest telescope of only 3¾ in aperture.

In 1952 a most interesting suggestion was made by Dr H.H. Nininger, an American astronomer who specialised in studies of meteorites. He believed that most of the lunar craters were of impact origin, and suggested that the Messier pair had been formed by a meteorite which sliced its way through a ridge, leaving a hole at either end and producing a true tunnel, which would connect the two. He wrote: 'If we assume a meteorite of large size travelling in an orbit nearly parallel to the Moon's surface at its point of incidence, entering to form Messier on the ascending slope of what we may call Tunnel Ridge, it would contact and ricochet from the solid sub-mantle at point X in the diagram, burning its way through the mantle again and emerging to form the Messier A hole.' He went on to suggest that as the meteorite emerged from the tunnel it would push compressed material in front of it, producing the famous double ray. He concluded: 'If and when the first explorers to the Moon have succeeded in landing there, perhaps they will find already prepared for them a shelter from small meteorites and, more important, from flying lunite slivers.'

We have come a long way since then; men have reached the Moon, small meteorites seem to be no hazard, and neither are there pieces of lunar surface flying around. Photographs of the Messier twins taken from close range show no evidence of a tunnel, and I fear that the idea must be abandoned. This is a pity; after all, a lunar tunnel would certainly have become a major tourist attraction when trips to the Moon can be made easily and often!

72   THE JEWEL BOX

There are some lovely sights in the sky: some of the planets, notably ringed Saturn; coloured double stars; gaseous nebulæ; and star clusters. I think that without doubt our 'top ten' of beautiful objects must include the Jewel Box, and it is a great pity that it is too far south to be seen from England.

The Jewel Box, known officially as Kappa Crucis or as NGC 4755, lies in the Southern Cross, which is a constellation just as familiar to Australians and New Zealanders as the Great Bear is to Britons. (NGC stands for the New General Catalogue of Clusters and Nebulæ, though it is no longer new; it was compiled by the Danish astronomer J.L.E. Dreyer over 90 years ago.) The Jewel Box contains dozens of stars, and you can just about see the cluster with the naked eye as a dim blur, though you need binoculars or, better, a telescope to see it properly. The

The Jewel Box cluster, around Kappa Crucis.

three brightest stars form a triangle. Inside the triangle is a line of three, one of which is very red. The name so often applied to the cluster was due to Sir John Herschel, who described it as being like 'a superb piece of jewellery'. It is often said that there are many stars in it of contrasting colours. This isn't strictly true, but that one red star, in the heart of the cluster, does stand out strikingly. Its true luminosity is about 16,000 times that of the Sun. This makes it as powerful as Betelgeux in Orion, but of course it is much further away. The distance of the Jewel Box is of the order of 7,700 light-years.

How big is it? Well, it seems that the 50 brightest members of the cluster are compressed into an area not more than 25 light-years in diameter, so that if we lived inside it we would have a gloriously starlit sky, with many stars bright enough to cast shadows. The most luminous members are at least 80,000 times more powerful than the Sun. Beyond the main cluster there are outlying parts which may stretch out for another 20 light-years or so.

The fact that the leading members are so luminous seems to show that the cluster is young by cosmical standards; remember that very powerful stars run through their life-cycles much more quickly than mild stars such as the Sun. The age of the Jewel Box may be only a few million years, and it is very likely that all its members condensed out of the same mass of gas and dust in space. And immediately south of the Jewel Box is a dark nebula, the Coal Sack, which hides any stars which lie beyond it.

If you have never seen the Jewel Box, I suggest that you go and look for it next time you happen to be in a southern country. You should have no difficulty finding it, and you will certainly not forget your first sight of this wonderful cluster with its bright blue stars and its red supergiant.

73 SPODE'S LAW: THE CASE OF ELPIS

In astronomy we have an unofficial law called Spode's Law. It states that if things *can* go wrong, they *do*; and if you don't want the sky to be cloudy, it *is*. Perhaps I may be forgiven if I digress briefly to give a typical case of Spode's Law in which I was the victim. . . .

It happened on June 14 1980, at 02:17 hours GMT; that is to say, in the early hours of the morning. It concerned Minor Planet number 59, Elpis, which had been discovered by the French astronomer Chacornac as long ago as 1860. It takes four-and-a-half years to go once round the Sun, and it lies well in the main asteroid belt, between the orbits of Mars and Jupiter. It never becomes brighter than magnitude 11, so that it is well out of binocular range; no telescope will show it as anything but a faint speck of light. Its diameter is a few tens of miles, but asteroid diameters are very difficult to measure because they look so small. We would very much like to know how big they really are.

At the Royal Greenwich Observatory, Gordon Taylor developed an ingenious method. Occasionally an asteroid must pass in front of a star, and hide or occult it. Then, obviously, the length of time for which the star is hidden tells us the size of the asteroid. Unfortunately occultations by asteroids do not happen often, and even when they do it is essential to be in exactly the right place at exactly the right time. For instance, if a minor planet occults a star as seen from England, it probably will not cause an occultation as seen from Scotland.

Gordon Taylor calculated that Asteroid 59, Elpis, would occult a dim star at about 02:17 hours GMT on June 14 1980. This was a great and rare opportunity, so he sent out circulars to all his observing team and made a special request for reports. I was one of the team, and I have a good 15-in reflector in my observatory at Selsey in Sussex. So I made preparations. On several successive nights I identified the star-field, and also Elpis, which was moving slowly and purposefully toward its target. All seemed well, and on the vital night the sky was beautifully clear.

At midnight my telescope was firmly guiding on the star. There it was; there was Elpis, and over periods of even a few minutes I could detect the shift. Then, to my annoyance, patchy clouds came up. I willed them to go away. They did. 01:00 hours:still all was well—Elpis was now nearing the star. I was frankly excited. This would be an important observation, and I was in just the right position.

More drifting cloud; the sky was rather light, but at 02:00 hours the field was still clear. 02:14 hours: less than three minutes to go—but Spode was watching. Up came the clouds. The star-field was blotted out; nothing remained. The vital moments passed; the occultation could not last for more than about two minutes at most. I realised, with deep depression, that I had missed it. And at 02:25 hours, when all was over, the clouds cleared; Elpis was past, and I will never know whether the occultation took place or not. It won't occur again in my lifetime.

It was a classic case of Spode's Law. Oh, well—I suppose that's Life!

74   A GREAT AMATEUR ASTRONOMER: J.P.M. PRENTICE

In 1981 the astronomical world was saddened by the death of a great observer, J.P. Manning Prentice, who will always be remembered both for his work in connection with meteors and for his discovery of the bright nova in Hercules in December 1934.

I well remember the nova. The first meeting of the British Astronomical Association which I attended was in December 1934; I was 11 years old, and very proud at having been elected a full member of the Association (I was, I think, the youngest member on record; as a personal note, I retired from the Presidency exactly half a century after my election). Prentice was there, and described how he had been taking a stroll after a spell of meteor-watching when he suddenly saw that there was something very odd about the region of Hercules and the head of Draco, the Dragon. A bright star had flared up where no naked-eye star had been before. At once he notified others of his discovery, and Nova Herculis was intensively studied.

A nova is not genuinely a new star. What happens is that in a pair of formerly faint stars, one undergoes a violent outburst which makes it flare up to many times its normal brilliance for a few days, weeks or months before fading back to obscurity. Nova Herculis became brighter than the Pole Star, and remained visible with the naked eye for many weeks. As it declined, it showed a vivid green colour very unusual in novæ (or in any other stars, for that matter). It is still visible, but it has now become a very dim telescopic object.

But Nova Herculis was incidental to Prentice's astronomical career (by profession he was a solicitor). He was an observer of meteors, and directed the

The Jodrell Bank telescope at night.

Meteor Section of the British Astronomical Association. At that time visual observations of meteors were all-important; it was the only way to find out how they moved, and how they were spread around in space. Prentice built up an energetic team, and his results were used by astronomers everywhere.

Then, after the war, came the development of radar, in which a pulse of energy is transmitted and 'bounced off' some solid body or equivalent; the echo can be received, and much can be learned about the object off which it has bounced. Meteor trails give radar echoes. Prentice went to Jodrell Bank in Cheshire and collaborated with Professor—now Sir Bernard—Lovell; they set up preliminary equipment, and found that the method worked well. That was to all intents and purposes the beginning of Jodrell Bank, now the site of the most famous radio telescope in the world.

Prentice was never a professional astronomer—he had no wish to be. But his work was universally recognised, he made fundamental advances in his own line of research, and he certainly ranks as one of the greatest of modern amateur astronomers.

75   DANGER FROM METEORITES?

Before the flight of Sputnik 1, the first artificial satellite, in October 1957 it was often thought that meteorites would prove to be a major hazard to space-travellers. It was even suggested that any vehicle incautious enough to venture beyond the Earth's protective screen of atmosphere would be treated rather in the manner of a cosmic coconut-shy, and would be battered to pieces. Fortunately this has not happened, and the danger is much less than had been feared. Yet there are plenty of solid bodies floating around in the Solar System, ranging from tiny meteors to large masses. So what are the chances of one of our space-craft being damaged in this way?

Apparently this is precisely what did happen on July 27 1983. According to an official Russian report, a tiny meteorite chipped a window of the Salyut 7 space-station, which had been in orbit for months. It made a loud 'crack' and left a ¼-in crater in the space-craft's window. Luckily the window was not actually broken. It was made up of two panels, each half-an-inch thick, to guard against this kind of eventuality. The Russian spokesman, Victor Blagov, said that it was nothing to worry about; an earlier space-station, Salyut 6, had been peppered with small craters.

This is all very well, and serious damage from a wandering missile may be improbable, but it cannot be entirely ruled out. Large meteorites have landed on the Earth, and no doubt upon the Moon and other planets also. And if a meteorite the size of, say, a dining-room table came into collision with a space-craft, there could be only one result. Moreover, taking evasive action would be difficult. Salyut 7's orbital speed round the Earth was around 17,000 mph, and it is quite out of the question to make a sudden swerve!

Yet on the whole, we may now be confident that the original fears were largely unfounded, which may well indicate that particles big enough and massive enough to be dangerous may be much less common than had been thought. Remember, too, that four space-craft—Pioneers 10 and 11, and Voyagers 1 and 2—have successfully passed through the asteroid belt, where the danger of collision might be expected to be much greater than elsewhere. I suppose that sooner or later we may have a tragedy caused by a collision between a space-craft and a sizable meteoroid, but it has not happened yet. Let us hope that it never does.

76   A DYING STAR

Some years ago I paid my first visit to the southern hemisphere. I found the sky unfamiliar; there are many splendid constellations which never rise above the British horizon—the Centaur, the Southern Cross and so on. But all in all, I was fascinated by a star which is no longer visible with the naked eye: Eta Carinæ, in the Keel of the now-dismembered constellation of Argo Navis, the Ship Argo.

I was using a powerful telescope belonging to a friend of mine, Christos Papadopoulos, Greek by birth but long resident in South Africa. We looked at Eta Carinæ, and had a superb view of it and the nebulosity surrounding it. Eta Carinæ does not look like an ordinary star at all; it resembles an orange blob. And certainly it is unique; for a period during the first half of the 18th century it was the most brilliant star in the sky apart from Sirius.

Eta Carinæ does not seem to fit into any known category. With a mass perhaps 120 times that of the Sun, and a surface temperature of 25,000 degrees centigrade, it is probably the most luminous star known. If its distance is 6,000 light-years (a reasonable estimate), then it is several million times more powerful than the Sun. Its total output may not have changed much since it shone so brightly, but less of the emitted energy is in the visible range. Also, it is associated with a superb nebula, and additional nebular material may have moved in front of it and masked it.

It was once suggested that Eta Carinæ might be a very young, unstable star, varying irregularly as it prepared to settle down to sober middle age; but this cannot be so, because no embryo star could be anything like so powerful. Then could it be a peculiar sort of nova? No; again the luminosity is too great, and so is the mass. Moreover, all novæ are binary systems, and there is no sign of a second component in the system of Eta Carinæ.

Now we have a new theory, developed by three American astronomers, N.R. Walborn, T.R. Gull and K. Davidson. They believe that instead of being a young star, Eta Carinæ is in an advanced state of senility, and that disaster looms ahead.

What they have done is to make spectroscopic studies of the wisps of nebular material ejected by Eta Carinæ in its glorious period (around 1835 to 1843), which have been moving outward and now lie ten seconds of arc away from Eta itself. First they used the powerful telescope at Cerro Tololo, in Chile, and looked for indications of oxygen. They found none, but there was abundant evidence of

The Eta Carinæ nebula (Cerro Tololo Observatory, Chile).

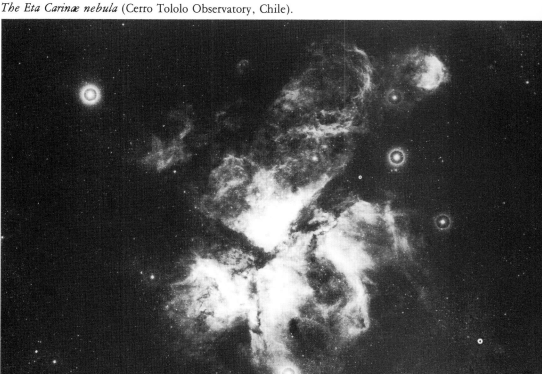

nitrogen. Next they called in an Earth satellite, the IUE (International Ultra-Violet Explorer) and hunted for carbon. Again the result was negative, but once more nitrogen was very much in evidence.

This was significant. In interstellar gas, oxygen is seven times as plentiful as nitrogen, and carbon four times as plentiful. Evidently the wisps from Eta Carinæ were decidedly out of the ordinary. But there could be a solution. Inside stars, at least very massive ones, nuclear reactions convert the original quantities of carbon and oxygen into nitrogen, so that eventually both carbon and oxygen are very much depleted. Eta Carinæ, then, is in an advanced stage of evolution, and has started to throw off shells of material. It is a stellar spendthrift, and is using up its reserves at a furious rate, so that before long (on the cosmical scale, that is to say) something cataclysmic will happen to it.

There are two possibilities. Eta Carinæ may explode as a supernova, in which case it will flare up and become a naked-eye object visible even in broad daylight for a few months. Yet with its exceptional mass, it may be a black hole candidate, so that when it starts to collapse it will simply go on doing so, until it can no longer emit light or anything else.

I invite you to take your choice. Of course, 'soon' is a relative term; Eta Carinæ may continue much as it is now for around 10,000 years yet, but it must die well before any dramatic change happens to the Sun. Meanwhile, astronomers are keeping a close watch on it. There is no reason why it should not brighten up again to become as magnificent as it was for a while in Victorian times.

77 T.J.J. SEE AND THE CRATERS OF MERCURY

One of the strangest characters in the history of modern astronomy is Thomas John Jefferson See. He was born in 1866 in Missouri, and worked successively at the Berlin, Yerkes and Lowell Observatories before joining the staff of the US Naval Observatory at Washington, DC in 1900. Here he made use of the fine 26-in refractor with which Asaph Hall had discovered the two tiny satellites of Mars, Phobos and Deimos, in 1877.

Though he had been largely concerned with measurements of double stars, See also paid attention to the planets. He was interested in Mercury, which is a very difficult object to see well; indeed, it was not until the flight of Mariner 10, in 1974, that we obtained reliable maps of its surface. But See, according to his reports, not only recorded markings, but also drew a large number of craters, and believed the surface to be not unlike that of the Moon apart from the absence of lunar-type 'seas'.

One of his drawings, made in June 1901, was remarkable by any standards. Wisely he observed Mercury when the planet was high in the sky; the presence of the Sun was preferable to the inevitably poor conditions when the planet was low down and could be found with the naked eye. When See made his sketch, Mercury's apparent diameter was only 6.6 seconds of arc, but conditions were excellent, and a high magnification could be used.

But was his work accurate? There can be no doubt that See was an oddity. To say that he was unpopular with his fellow astronomers is to put it mildly, and many of the Lowell Observatory staff left because they could not tolerate

The region of Mercury's south pole, taken by Mariner 10 during its second flyby on September 21 1974. The range was 33,850 miles and the area shown is 285 × 217 miles.

him—which is why See himself was forced to leave in 1898. The general opinion was voiced by A.E. Douglass, one of Lowell's early and most senior assistants. Writing to a colleague, he claimed that See showed 'evidence of mental degradation', and went on: 'Personally, I have never had such aversion to a man or beast or reptile or anything disgusting as I have had to him. The moment he leaves town will be one of vast and intense relief, and I never want to see him again. If he comes back, I will have him kicked out of town.'

See did not come back, and it was, of course, after he left the Lowell Observatory that he made his drawing of Mercurian craters. The fact that his claim has been generally rejected may well be because of his personality, and it has been pointed out that if he had faked the drawing he would presumably have included some plains of lunar type—which he did not. So the question remains open; all we can say is that there is at least a chance that T.J.J. See saw craters on Mercury more than 70 years before the surface was mapped in detail from Mariner 10.

78 THE REMARKABLE 'KID'

Close beside Capella, the brilliant yellow star which is almost overhead as seen from Britain during the winter evenings, lies a triangle of much fainter stars, nicknamed the Hædi or 'Kids'. One of these—Epsilon Aurigæ, at the apex of the triangle—looks ordinary enough, but has proved to be one of the most extraordinary objects known to us.

Normally it shines as a star of the third magnitude, so that it is very easy to see with the naked eye, and is very slightly brighter than the second brightest Kid, Eta Aurigæ, which is 370 light-years away and about 600 times as luminous as the Sun. But Epsilon Aurigæ is not constant. Every 27 years it fades slowly down by almost a magnitude, remaining at minimum for a year before slowly regaining its lost lustre.

It is not a single star. The visible object is a yellowish supergiant, around 60,000 times as powerful as the Sun and about 3,200 light-years away. Associated with it is an invisible companion, and it is this companion which passes in front of the supergiant every 27 years, cutting off part of its light. But what is this companion? It is utterly invisible; it does not radiate even at infra-red wavelengths; but we can estimate its mass, which proves to be about 20 times that of the Sun (as against 35 Suns for the supergiant). Now, few stars are as massive as this, and the secondary would certainly be visible if it were a normal object. We have a problem to solve.

The first idea was that the companion was a very young star, not yet hot enough to shine. If so, then the duration of the eclipse would give a clue as to its diameter—around 2,000 million miles, so that it could swallow up the orbits of all the planets in the Solar System out to beyond Saturn. However, even at mid-eclipse the supergiant can still be seen, and we can hardly visualise a young star which is transparent.

The next suggestion was that we might be dealing with a black hole; that is to say, an old collapsed star which is pulling so powerfully that not even light can escape from it. But with a black hole, material just about to be sucked in would be so strongly heated that it would emit high-energy radiation, and nothing of the sort has been detected.

Could the secondary be a disk-shaped cloud, so placed that when passing in front of the supergiant it bisects it, leaving some of the bright star visible? Again there are difficulties. Such a cloud would be unstable, and would also be detectable in infra-red.

It now seems more likely that the companion is a smallish, hot star surrounded by a shell or disk of gas, and that it is this disk, so heated that it has become opaque, which causes the eclipses. We might expect to find a certain amount of short-wave radiation, and there are indications that this is so, but it is not easy to believe in a shell as extensive as the eclipsing companion would have to be.

The latest eclipse is just over. It began in July 1982, was total from January 1983 to January 1984, and ended in June 1984. Astronomers all over the world were paying close attention, because nothing more will happen for the next 27 years; but though all the most modern techniques were used, we have to admit that the mystery of Epsilon Aurigæ has yet to be solved. It is indeed a remarkable 'Kid'.

Facing page *Position and theories of Epsilon Aurigæ.*

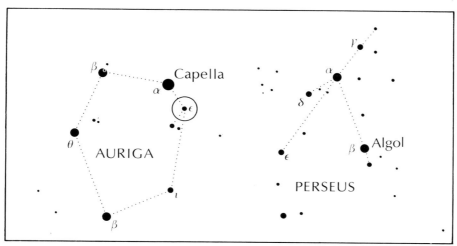

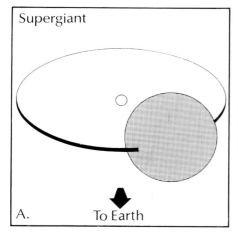

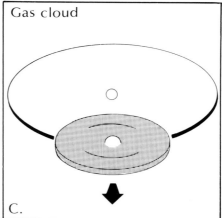

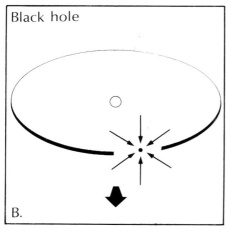

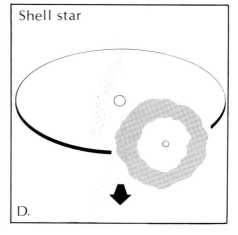

79 KOZYREV AND THE MOON-GLOW

The Soviet Union has produced many famous astronomers during the past half-century. Nikolai Alexandrovich Kozyrev may not have been in the first rank, but he was certainly responsible for an observation which had far-reaching results.

Kozyrev was born in February 1908, and published his first astronomical paper at the age of 17. He turned his attention to problems of the stars, but his career was interrupted in 1937 when he was arrested during Stalin's reign of terror and thrown into prison. He was not released for 11 years, and it must have been a terrible time, though no charges were ever brought against him. When he was finally freed, he went to the Pulkovo Observatory, near Leningrad, and began to study the surfaces of the Moon and planets. It was around this time that I began to correspond with him, though we did not actually meet face to face until 1952.

Kozyrev used the excellent 50-in reflector at the Crimean Astrophysical Observatory for his lunar work, and spent more time there than in Leningrad. Around that time lunar observers in Britain were searching for what are termed Lunar Transient Phenomena—that is to say, very slight outbreaks on the Moon's surface, due probably to gas and dust being sent out from beneath the crust. One area where such things had been reported was the floor of the large crater Alphonsus, a huge ringed structure over 80 miles in diameter. I suggested to

Below left *The Mare Humorum on the Moon, as I photographed it with my 15-in reflector in 1981. The large crater at left (N) boundary is Gassendi, 75 miles in diameter.*

Below right *The Moon through the Mount Wilson 100-in reflector. Ptolemy is the large crater at the bottom; above it are Alphonsus and Arzachel. The Fra Mauro area, where Apollo 14 landed, is to the lower right. The Straight Wall is nearly central in the picture.*

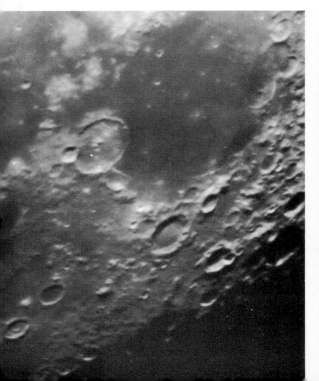

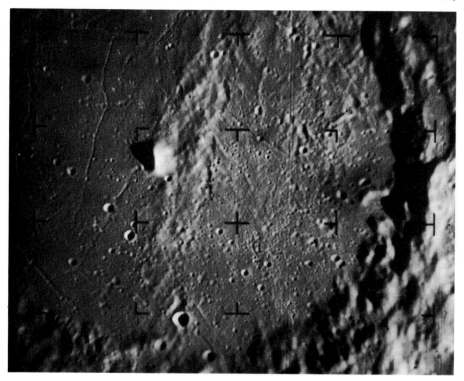

Alphonsus, photographed from the crash-lander Ranger 9 just before impact on March 21 1965. Up to that time, this was the most detailed view of the crater ever obtained.

Kozyrev that this was a region well worth watching. And at 01:00 hours GMT on November 3 1958, Kozyrev saw that the central peak of Alphonsus had become blurred, and was apparently engulfed in a reddish cloud. Between 03:00 and 03:30 hours GMT the central peak became abnormally bright, and Kozyrev was able to obtain photographic confirmation. The outbreak ended at about 03:45 hours.

This was the first observation of such an event made photographically by a professional astronomer. Kozyrev's explanation does admittedly seem rather dubious; he believed that the peak was 'being illuminated by the Sun through the dust and ash thrown up by the eruption', and that there had been a rise in temperature of 2,000°F. Later observations of similar outbreaks on the Moon do not confirm ideas of this kind, but at least there can be no doubt that outbreaks of some kind or other do occur. The Moon is not entirely inert.

Kozyrev continued his lunar work, though he never again saw an event as spectacular as that of 1958. He also produced some new and revolutionary theories involving space and time, which are highly controversial. But whether right or wrong, Nikolai Kozyrev himself was a most sincere and pleasant man, and a good friend. His death in February 1983 was deeply regretted by his many colleagues all over the world.

80   THE OCEANS OF TITAN

Saturn, the planet with the glorious system of rings, is possibly the most beautiful object in the entire sky. It also has a fascinating family of satellites, of which perhaps the most intriguing is Titan, discovered by the Dutch astronomer Christiaan Huygens as long ago as the year 1655.

Titan is a big world—larger than our Moon, slightly larger than the planet Mercury, so that it is visible with a small telescope as a starlike point. On November 12 1980 the probe Voyager 1 bypassed it at only about 4,000 miles, but the pictures showed nothing more than a layer of orange cloud, because Titan has a thick atmosphere which hides its surface permanently and completely. The atmosphere proved to be made up chiefly of nitrogen (the gas which makes up 78 per cent of our own air), which was a surprise. There was also about 6 per cent of the gas methane, which is a hydrogen compound.

Voyager 1 indicated that the ground pressure of the atmosphere was about one-and-a-half times that of the Earth's air at sea-level. The temperature of the surface was about – 168 degrees centigrade. Now, this is between the melting point of methane (– 182 degrees) and its boiling point (– 155 degrees). The atmospheric methane vapour presumably rises from a source on the surface, and the temperature measures indicate that on Titan's surface the methane is likely to be in liquid form.

In America, Drs Carl Sagan and Stanley Dermott have put forward a fascinating theory. They assume that on Titan there is a methane ocean. Saturn will produce strong tidal effects on this ocean, each tide amounting to something like 30 ft per revolution (Titan takes just under 16 days to complete one orbit round Saturn). One might expect friction of the tides against the solid surface to move Titan into a path in which the tides are reduced to zero. However, this has not happened, and so the friction must be very slight. With a shallow ocean, of course, the effects would be comparatively strong.

This seems to leave only two possibilities. Either Titan has no liquid methane on its surface at all, or else there must be a methane sea which is very deep. With no surface liquid, the presence of methane vapour in the atmosphere would be hard to explain. Accordingly, Sagan and Dermott conclude that the ocean has a depth of at least 1,200 ft, which provides a reservoir of natural gas 200 times greater than on the Earth.

If this is correct, then Titan is a stranger world than we expected even after the Voyager passes. Unfortunately there seems little chance of obtaining proof in the near future, because no more Saturn probes are planned as yet. All we can say is that there is a very good chance that the surface of Titan is covered with a deep ocean of methane, which makes it unique in the Solar System. One may even dare to suggest that future astronauts going there will need not an ordinary probe vehicle, but something more in the nature of a cosmic submarine!

81   ZANIAH

Virgo, the Virgin, is one of the more conspicuous of the Zodiacal constellations, and contains the first-magnitude star Spica. There are also some interesting telescopic objects, notably the fine binary Arich or Gamma Virginis, and a wealth of faint galaxies.

Zaniah, or Eta Virginis, makes up part of the familiar 'Y' pattern, but is not prominent; its magnitude is exactly 4, which means that it is a very easy naked-eye object on a dark night, but is apt to be swamped when close to the Moon. It is white; its surface is hotter than that of the Sun, at about 11,000 degrees centigrade; and it is 104 light-years away, so that we are now seeing it as it used to be in Victorian times. It is approximately 21 times as luminous as the Sun.

Zaniah has a somewhat unusual history. The first proper star-catalogue which has come down to us was drawn up in the second century AD by Ptolemy of Alexandria. Ptolemy naturally included Zaniah, but he gave its magnitude as 3, and made it equal to Arich, in the base of the 'Y' bowl of Virgo. Arich is still of the third magnitude, so that it is not only brighter than Zaniah, but very much brighter.

Around the year 903 an Arab astronomer named Al-Sûfi produced another excellent star-catalogue. He too made Zaniah of magnitude 3, and equal to Arich. Much later—in the 19th century—the German astronomer Eduard Heis made it 3½, and now, as we have noted, it is down to 4. It doesn't seem to have altered much for many years.

If Ptolemy and Al-Sûfi were right, Zaniah used to be considerably brighter than it is now. On the other hand it is not the sort of star which would be expected to show steady fading. It should be completely stable, and unlikely to alter over periods of millions of years. So what is the answer? As with other stars which have allegedly faded or brightened up since ancient times (Castor is a classic example; so is Megrez in the Great Bear), it seems most likely that the old astronomers were either mistaken or misreported. With Zaniah we do have the testimony of Heis, who was an extremely accurate observer and was not likely to make an error of half a magnitude. So although any real change in Zaniah over the past 2,000 years is frankly very improbable, a slight doubt remains.

82   THE FLIGHT OF LUNIK 3

Look at the Moon, even with the naked eye, and you will see bright and dark patches. The dark patches are called seas, even though there has never been any water in them; they are old lava-plains, formed when the Moon was an active world thousands of millions of years ago. Yet from Earth we can see only one hemisphere. The other side is always turned away from us—and this brings me on to our first views of the far side, in October 1959.

The Moon takes 27.3 days to go once round the Earth. (To be pedantic, the Earth and Moon move together round their common centre of gravity, but since this point lies inside the Earth's globe the simple statement is good enough for most purposes.) The Moon also spins on its axis in 27.3 days, which is why it always keeps the same face turned in our direction. There is no mystery about this curious behaviour; tidal friction over the ages has been responsible. Actually, some minor 'wobblings' called librations mean that we can examine a total of 59 per cent of the whole surface, though of course never more than 50 per cent at any

122

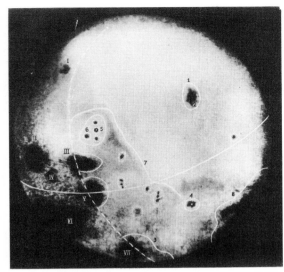

Left *Early picture of the Moon's far side from Lunik 3 in October 1959. The features are: 1 Mare Moscoviense; 2 Bay of Astronauts; 3 Part of Mare Australe; 4 Tsiolkovski; 5 Lomonosov; 6 Joliot-Curie; 7 Sovietsky Range; 8 Sea of Mechta. I Mare Humboldtianum; II Mare Crisium; III Mare Marginis; IV Mare Undarum; V Mare Smythii; VI Mare Foecunditatis; VII Mare Australe. The 'Sovietsky Range' was originally thought to be mountains but has subsequently been discovered to be nothing more than a bright ray.*

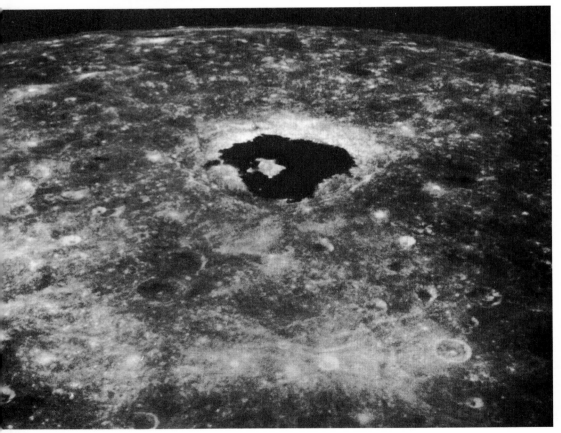

Below left *The remarkable crater Tsiolkovsky, on the Moon's far side, first identified from Lunik 3 in 1959. It is 150 miles in diameter, with a dark lava floor and a central peak.*

Right *The crater Giordano Bruno, on the Moon's far side. It is the bright ray-centre near the top.*

Below *Copernicus, photographed from Orbiter 2 on November 23 1966. At the time this was called 'the picture of the century'.*

one time. The remaining 41 per cent is permanently out of view. (Note, however, that the hidden area is not permanently dark. It is always turned away from the Earth, but not from the Sun.)

Before 1959 everyone wanted to know what lay on the hidden side of the Moon. Certainly I did; I had spent many years in charting the edge of the Moon as seen from Earth, using the telescopes at my own observatory in Sussex as well as some larger instruments elsewhere. In 1958 I had a letter from the Soviet Academy of Sciences in Moscow: would I please send them all my charts of the Moon's edge, as well as the observations I had not published yet? Of course I did so, and there the matter rested until October 1959.

On October 4 of that year (exactly two years after the launching of the first artificial satellite, Sputnik 1), the Russians sent up Lunik 3. This was an unmanned rocket vehicle; it went round the Moon, took pictures of the far side, and then swung back towards us. On October 26 it transmitted its pictures. I was in a BBC television studio when they came through; I had just started a *Sky at Night* programme—and then, suddenly, the far side of the Moon flashed on to the screen. It was a thrilling moment. Fortunately I was able to identify one of the familiar lunar features, seen under reverse lighting, and I was able to give what I hope was an intelligible commentary. By present-day standards the pictures were blurred and lacking in detail; by 1959 standards they were superb. At least we had final proof that the far side of the Moon is as barren as the side we have always known.

Today the whole of the Moon has been mapped, but that moment on October 26 1959 was something I will never forget—and I am also glad to say that my charts, some of which the Russians used to link the far-side with the near-side maps of the Moon, proved to be reasonably accurate.

83   THE RING OF NEPTUNE

The glorious ring-system of Saturn has been known since the early days of telescopic observations; Galileo saw it, though admittedly he did not know what it was. In modern times two more of the giant planets have been found to be ringed. Voyager photographs have revealed a thin, dark ring round Jupiter, and in 1977 a whole system of rings round Uranus was detected in a remarkable way. Uranus passed in front of a star, hiding or occulting it. Both before and after the occultation, the star gave a series of regular 'winks', so that evidently it was being briefly obscured by Uranian rings—and since then there has been visual confirmation. This leaves Neptune, which, as you will recall, was discovered in 1846 by two German astronomers, Galle and D'Arrest, on the basis of mathematical calculations made by Urbain Le Verrier in Paris.

Actually, the English mathematician John Couch Adams had made similar calculations, and if his work had been acted upon the honour of the discovery would have gone to England, not to France. A search had been made, but in a rather desultory fashion and, moreover, too late. The observer concerned was James Challis, of the Cambridge Observatory, who will always be remembered as the man who did *not* discover Neptune even though all the information had been put into his hands.

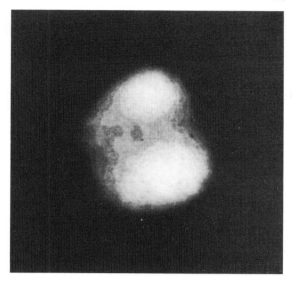

Neptune, from the Catalina Observatory in Arizona. Dark features are shown in the planet's atmosphere.

But for a chain of misfortunes William Lassell, a leading British amateur, would have joined in the search, and he would undoubtedly have been much more energetic than Challis was. To provide Britain with at least some of the glory, Lassell did discover Triton, Neptune's large satellite, only a few weeks after the identification of Neptune itself.

Using his fine 24-in reflector, Lassell also reported the existence of a ring, and in October 1846 even described Neptune as 'very like Saturn as seen with a small telescope and a low power'. Challis confirmed the existence of the ring, but other observers were dubious. After Lassell moved his 24-in reflector to the clearer skies of Malta he was unable to recover the ring, and finally retracted, dismissing the alleged ring as nothing more than an optical effect.

In this he was, of course, right. No telescopes of the period could have shown a Neptunian ring even if it existed. But as Jupiter, Saturn and Uranus all have ring systems, why should Neptune lack one?

As long ago as 1979 I gave a brief paper in London in which I put forward my view that Neptune will prove to be ringless. This is because Triton, its large satellite, moves round the planet in a retrograde direction—that is to say, in the opposite sense to that in which Neptune is rotating. I believe that this will make for unstable conditions, so that no ring could form.

We should know for sure in 1989, if Voyager 2 is still working as it flies past Neptune. If no ring exists, I will recall that I said as much. If it does, I will forget that I ever made a comment about it!

84 THE BARONESS AND THE SUPERNOVA

On the evening of August 22 1885 a Hungarian lady, the Baroness de Podmaniczky, was giving a house-party. One of her guests was an astronomer, Dr

Messier 31, the Great Spiral in Andromeda. Unfortunately, the Baroness' supernova appeared in 1885, before regular photography had been developed in astronomy.

de Kövesligethy, and since the Baroness herself was mildly interested in the stars she owned a 3 ½-in refractor. Taking it on to the lawn, she and some of her guests 'looked around'. One object in view was Messier 31, the Great Spiral in Andromeda, now known to be an independent galaxy larger than our own, lying at a distance of rather over two million light-years.

Gazing through the telescope, the Baroness remarked that she could see a star in the midst of the nebula. Dr de Kövesligethy agreed, but believed that the appearance was due to the presence of the Moon, which drowned the fainter part of the nebula; and neither thought much more about it—until they heard that the star had been seen elsewhere, and was certainly unusual.

In fact, Professor Ludovic Gully, at Rouen in France, had noted the star on August 17, but put it down to a fault in his telescope. But on August 20 the star was seen by Dr Hartwig, who was on the staff of the Dorpat Observatory in Estonia (then, as now, controlled by Russia). Hartwig did not fall into the same trap. He realised at once that the star was new, but the Director of the Dorpat Observatory refused to allow him to announce it until the discovery had been confirmed in a moonless sky. Hartwig did write to Kiel, the centre of astronomical information, but his letter never arrived. Postal services in 1885 were just as unreliable as they are today.

Max Wolf, at Heidelberg, saw the star on August 25 and 27, but he too put it down to the effect of moonlight. Later, the Irish astronomer Isaac Ward claimed to have seen the star on August 19, but gave its magnitude as 9 ½, whereas the other observations make it very much brighter. As soon as the Moon had disappeared, Hartwig re-observed the star, and this time there was no room for doubt.

We now know that the star—officially listed as S Andromedæ—was a supernova: a colossal stellar outburst, so that a formerly faint star blazed up until it was shining at least 15 million times as brilliantly as the Sun. It soon faded, and by mid-September had disappeared. Certainly it will never be seen again, and it remains the only supernova to have been seen in the Andromeda Spiral, though there have been plenty of ordinary novæ.

Even then its nature was not appreciated, and some eminent astronomers, such as Trouvelot in France, believed it to be an ordinary nova in the foreground. It is a great pity that it appeared before modern methods of observation could be used. But at all events it is unique in one respect: it remains, and probably always will remain, the only supernova to have been discovered by a Hungarian baroness.

85 INTERGALACTIC TRAMPS

For many years now we have known that our Galaxy is a flattened system, with a central bulge; the Sun, with the Earth and the other members of its family, lies perhaps 33,000 light-years from the galactic nucleus (some astronomers believe this to be a slight over-estimate), and the diameter of the Galaxy is of the order of 100,000 light-years, though again there is some uncertainty. Surrounding the main Galaxy is the 'halo', made up of stars together with the huge, symmetrical systems which we call globular clusters. Beyond, there is a vast gap before we come to the next galaxy.

But is this gap completely empty? Undoubtedly it contains very tenuous material—far more thinly-spread than the atoms in the best vacuum we can create in our laboratories. Today it seems that there may be other objects as well.

Far away from us—16,000 light-years or so—there is a globular cluster known as NGC 5694. It is not conspicuous; it appears as a dim blur of the tenth magnitude.

Messier 13, the globular cluster in Hercules which is just visible to the naked eye (Mount Wilson Observatory).

What makes it so special is that according to measurements of its velocity and the direction in which it is moving, it seems to be on its way out of the Galaxy altogether. If so, it will move away until it has left us behind, and it will never come back. It will be a wanderer between the galaxies: an Intergalactic Tramp.

Of course, there must again be a degree of uncertainty, but all the evidence indicates that this really will happen. If so, then there may well be other globular clusters which also have escaped from the Galaxy. Populous though they are (some of them contain up to one million stars), globular clusters are much less luminous than average galaxies, so that at great distances we could not see them. Therefore, we have no real idea of how many 'tramps' there are.

We can take matters a stage further. If a globular cluster can escape, then, presumably, so can a single star. Once beyond the galactic halo, a star would be beyond the limit of detection with the instruments at our disposal today, particularly if it were a star no more luminous than the Sun.

We can visualise the sky as seen from a planet moving round such a star. The night sky would be virtually blank; nothing would be seen apart from dim glows in the extreme distance, apart of course from any other planets in the system of the 'tramp' itself. It would seem decidedly lonely, and I think we must be grateful that our Sun is not a solitary wanderer in the space between the galaxies.

86 THE AMAZING DR LESCARBAULT

It has been said that one of the rudest men who has ever lived was the great French astronomer Urbain Jean Joseph Le Verrier. Of his ability there could be no doubt at all; it was he who provided the calculations which led to the first identification of the planet Neptune in 1846, but he was certainly not popular. One of his colleagues commented that although he might not be the most detestable man in France, he was certainly the most detested. He was forced to resign the Directorship of the Paris Observatory in 1870 because of his 'irritability', though he was reinstated when his successor, Delaunay, was drowned in a boating accident.

Le Verrier had tracked down Neptune because of its effects upon the movements of Uranus. He then made similar calculations with regard to Mercury, and came to the conclusion that it was being perturbed by a planet still closer to the Sun. But how could this planet be seen? It would be drowned by the Sun's rays; the only hope would be to catch it as it passed in transit across the solar disk, directly between the Sun and the Earth.

In December 1859 he received a letter from a French country doctor, Edmond Lescarbault, who lived in the small town of Orgères. On March 29 1859, Lescarbault wrote that he had seen a small, round black spot pass slowly across the Sun's face. Could this have been the expected planet? Le Verrier was incredulous; still, he was interested enough to make the journey to Orgères and interview Lescarbault in person. According to the detailed account written by R.M. Baum,[*] Le Verrier began by leaving the doctor in no doubt about his scepticism. 'It is then you, sir, who pretend to have observed the intra-Mercurial planet, and who have

[*] R.M. Baum: 'Le Verrier and the Lost Planet', *Yearbook of Astronomy*, London 1982.

Urbain J.J. Le Verrier.　　　　*The search for Vulcan.*

committed the grave offence of keeping your observation secret for nine months. I warn you that I have come here with the intention of doing justice to your pretensions, and of demonstrating either that you have been dishonest or deceived. Tell me, then, unequivocally, what you have seen.'

One may well imagine that Lescarbault was taken aback, but he explained what he had observed. Le Verrier became even more astonished at the doctor's 'timekeeper', which was an ancient watch devoid of its second hand; seconds were timed by means of a pendulum made up of an ivory ball hung by a silk thread on to a nail in the wall. His method of recording his observations was equally unorthodox. Paper was scarce; Lescarbault was the village carpenter as well as being a doctor, and he wrote his observations down on planks of wood, planing them off when he had no further use for them.

In view of all this, it is surprising that Le Verrier came away converted. Yes, Lescarbault had seen the planet in transit; on January 2 1860 Le Verrier gave a formal account of the discovery to the French Academy of Sciences, and the planet was even given a name—Vulcan. Le Verrier calculated that its mean distance from the Sun was 13 million miles, and its period 19 days 17 hours.

But doubts soon crept in. Liais, a French astronomer living in Brazil, had been observing the Sun at the same time as Lescarbault, and had seen nothing at all. No further transits were observed, though Le Verrier was able to predict them on the basis of his computed orbit. But Le Verrier himself did not waver; as recently as

1874 he stated that the existence of Vulcan was 'beyond doubt'. He died in 1877.

Since then efforts have been made to detect Vulcan during total solar eclipses, when the sky is darkened, but the results have been as unconvincing as they are contradictory. Moreover, the irregularities in Mercury's motion have been cleared up by the theory of relativity without the need for introducing an extra planet. Vulcan does not exist, and never did. All the same, the country doctor of Orgères has his modest place in the history of astronomy!

87   CHIRON, THE MINI-PLANET

In 1977 an exciting discovery was made by a famous American astronomer, Charles Kowal, who was working at the Palomar Observatory in California. Kowal was making a systematic search for new comets when he suddenly came upon an object which was most certainly not an ordinary comet. With his colleague Tom Gehrels, he started checking earlier photographs of the same area—and there was the object. Working out the orbit, Kowal found that the object moved around the Sun at a mean distance greater than that of Saturn, so that it spent much of its time in the region between the orbits of Saturn and Uranus.

This was completely unexpected. Chiron, as the object was named (in honour of the wise centaur who taught Jason and the other Argonauts of mythology) seemed to be in a class of its own. Was it an asteroid or minor planet? It seemed rather large, with a diameter of several hundred miles; but after all Ceres, the largest asteroid, is 600 miles across. It looked exactly like an asteroid, and was duly given a number, 2060. But as Kowal commented, the region between Saturn and Uranus was the very last place where an asteroid would be expected.

Chiron's revolution period is 50.7 years, and its distance from the Sun ranges between 794 million miles and 1,756 million miles. One-sixth of its orbit lies closer-in than Saturn, but at its aphelion (greatest distance from the Sun) it may

Below left and right *The Palomar 200-in reflector, and its dome, as I saw them in 1976. The reflector looks the same now, although it is used with electronic equipment instead of photographic plates.*

be further out than Uranus. Its orbit is inclined by 6.9 degrees to the plane of the ecliptic, and the orbital eccentricity is 0.38, greater than that of any planet, but much less than those of most comets and some 'normal' asteroids.

Chiron has been traced on photographic plates taken as long ago as 1895, but of course it is very faint, and it is not surprising that it escaped detection for so long. When Kowal discovered it the magnitude was 18. When it reaches perihelion, in 1996, the magnitude will rise to 15, but this is still near the limit for telescopes of the size used by most amateur astronomers. There is no chance of a collision with either Saturn or Uranus, but it has been calculated that in the year 1664 BC it approached Saturn to a distance of 10 million miles, which is not much greater than the distance between Saturn and its outermost satellite, Phœbe.

Chiron appears to be darkish, in which case it is not an icy world such as many of the satellites of the giant planets. There have been suggestions that its orbit is unstable, so that eventually it may be thrown out of the Solar System altogether. Meanwhile, its precise nature remains uncertain. Kowal himself summed up the situation perfectly when he commented that Chiron was—'well, just Chiron!'

88   THE PIGEON PROBLEM

In the mid-1960s the American astronomer R.H. Dicke was carrying out theoretical investigations about the early history of the universe. He assumed that the universe originated some 15,000 to 20,000 million years ago in what is commonly called the Big Bang. All the material appeared at the same moment, and at first the temperature was very high indeed—a great many millions of degrees. Gradually, as the universe expanded, the overall temperature fell. Today, Dicke calculated, it should be around – 270 degrees centigrade or – 454 degrees Fahrenheit, which is only about three degrees above 'absolute zero'—that is to say, the coldest temperature possible. In this case there should be a weak background of radiation, with wavelengths between about one mm and one m, coming from all directions. This is the 'microwave' part of the total range of wavelengths—just beyond the infra-red.

Dicke began considering how to build apparatus suitable for detecting this microwave radiation. But unknown to him, not very far away, two Bell Telephone Laboratories scientists, Arno Penzias and Robert Wilson, had built precisely this sort of apparatus— for a completely different reason. Their main aerial was shaped like a horn, and was known as the 'horn antenna'. To their surprise, Penzias and Wilson detected weak microwave radiations which they could not identify.

Could they be cosmic? At first the probable answer seemed to be 'No', and there was a much more homely explanation to hand. Pigeons were plentiful in that part of the United States, and they are not particular in their habits. They settled on and in the horn, and, naturally enough, left their marks in no uncertain manner. Penzias and Wilson were convinced that what they were detecting with their sophisticated apparatus was simply the radiation given off by pigeon droppings!

It is rather difficult to argue with a pigeon, and the only solution was to make sure that the apparatus was completely free of droppings and then put up screens to ensure that the pigeons could not return. This took some time and effort, but eventually it was done. The microwave background was still there.

At about this time Dicke heard about the experiments, and realised that the microwave background was exactly what he had been expecting. Once the pigeons had been eliminated, the real cause of the radiation became clear.

Probably this so-called 3-degree radiation is the most important single argument in favour of the Big Bang theory of the origin of the universe, and it is rather amusing to look back to the time when it was put down to unco-operative pigeons.

89 PERSONAL MISADVENTURES

Since I first became seriously interested in astronomy, which was at the early age of six, I suppose I have made more mistakes than most people. Some of the misadventures have been my fault; others have not.

One episode, which I remember well, occurred in Northern Ireland in 1965. For a period of three years I was Director of the new Armagh Planetarium, and I used to make full use of the fine 10-in refractor at the Observatory. The dome was, frankly, somewhat antiquated, and to open the slit one had to climb a pair of rickety steps and pull the slit back. This was more awkward for me than for most people, because during the war I had my right wrist damaged and it lacks strength (which is why I have no really powerful forehand smash when playing tennis).

Therefore I had to operate the slit with my left hand. At about three o'clock one morning, after an observing run, I mounted the steps, pulled hard—and shut myself in the slit; my head was protruding from the dome and my legs dangling in the air. Trying to release myself I kicked the steps away. I must have been there for half an hour before, mercifully, someone heard me and came to the rescue.

Much earlier, I had been timing an occultation of a star by the Moon—a timing I was particularly anxious to secure. All went well. Unfortunately I stepped backwards, fell over the cat, and my hands closed convulsively on my stop-watch, re-starting it—so I will never know whether that timing was accurate or not.

Domes can cause trouble. My 15-in reflector is on a fork mounting with clock drive, but the dome does not move with the telescope; it has to be adjusted by hand. I was once making a long series of transit-times of surface features on Jupiter when I found that I was losing 'light'. The sky seemed clear; I went back to the telescope—no improvement. I then checked all the optics, without result. Only after about 20 minutes did I realise that the telescope had moved so much that the dome was now in front of it, and I was trying to look through solid metal.

Television, of course, provides plenty of scope for disaster. The fiftieth *Sky at Night* programme (in 1959!) was not meant to be a comedy show, but it turned out that way. We were 'live', and we had cameras fixed to a 24-in telescope to show the Moon and Mars direct. You can guess what happened. Every time we had the telescope correctly aimed, clouds intervened. After half a dozen attempts our time ran out. Minutes later, the whole sky was beautifully clear. . . .

Finally, there was the total solar eclipse of 1961. I was carrying out a live commentary from the top of a mountain in Jugoslavia. Unknown to me, the Jugoslav producer was a man with ideas. We had some mountain oxen up there, and the producer's idea was to show them as soon as the sky darkened and the eclipse became total; he had been told that under such conditions animals imagine that night has fallen, and go to sleep. At totality, he duly trained the cameras onto the oxen. Just to make sure that everyone could see them nicely, he floodlit them. The experiment was not a success. Still, we did see totality, so I suppose things might have been worse.

Totality. This was as I photographed it in 1983 from Java. The corona was well seen.

90　47 TUCANÆ

Globular clusters are among the more remarkable members of the Galaxy. They are symmetrical, and may contain up to a million stars. One of them, Messier 13 in the constellation of Hercules, is visible with the naked eye as a hazy patch; a large telescope will resolve it into stars almost to its centre. However, Messier 13 is far inferior to two globular clusters which are too far south to be seen from England. One is Omega Centauri; the other is 47 Tucanæ in the otherwise unremarkable constellation of Tucana, the Toucan.

Globular clusters lie around the edge of the main Galaxy, and it has been said that they provide a sort of outer framework to it. Some of the stars contained in them are variable in brightness. Short-period variables betray their distances by the way in which they behave, and this of course provides an invaluable clue to the distances of the globulars themselves. (It was in this way that the American astronomer Harlow Shapley, more than 60 years ago, first gave a reliable measurement of the size of the Galaxy.) Unfortunately 47 Tucanæ does not contain many short-period variables, and its distance is somewhat uncertain, but it seems to lie between 16,000 and 20,000 light-years from us.

Great though this distance may seem, 47 Tucanæ is one of the nearest of the globular clusters. Small telescopes will resolve its outer parts into stars; with large instruments even the centre is resolvable. It looks as though the stars are so crowded that they are in imminent danger of colliding with each other. Actually this is not so; even in the middle of a globular cluster the stars are still, on average, light-weeks apart, but from a planet in such a system the sky would be ablaze. Many stars would cast shadows, and a large number of them would be red, because globular clusters are very old systems, and their leading stars have evolved into the red giant stage.

Omega Centauri is closer, larger and brighter than 47 Tucanæ, but seen through a telescope it is probably less imposing, because it more than fills the field when a reasonably high magnification is used. To me, 47 Tucanæ is the most beautiful of all the globular clusters; and tiny though it may look, it is really more than 200 light-years across. European astronomers never cease to regret that it lies so far south in the sky.

Left *47 Tucanae, the brightest globular cluster in the sky apart from Omega Centauri, photographed by Alan Gilmour with the 24-in reflector at Mount John, New Zealand.*

Right *The pulsar in the Crab Nebula, seen at the junction of the white pointers.*

91   THE FURTHEST PULSAR

Of the external galaxies, two of the nearest are the Clouds of Magellan, visible only from latitudes well south of Europe. The large Cloud is about 190,000 light-years from us, and contains objects of all kinds: stars, star-clusters, nebulæ and all the rest. We now know that it also contains at least one detectable pulsar.

When a very massive star uses up its last reserves of energy, it collapses, and it may suffer a sudden, tremendous implosion—the opposite of an explosion. Most of its material is hurled away into space in what we call a supernova outburst, and in some cases (not all) all that is left is a patch of expanding gas, in the midst of which is a very small, super-dense object called a pulsar. It is made up of neutrons, particles with no electrical charge, and it is so dense that a cupful of it would weigh millions of millions of tons. It is spinning round very rapidly, and since it has a strong magnetic field it sends out pulses of radio radiation—hence the name pulsar.

The first pulsar was discovered in 1968 by Dr Jocelyn Bell-Burnell at Cambridge, in England. Others followed, and the grand total now runs into hundreds, though only two normal pulsars have been identified with optical objects. One is in the Crab Nebula in Taurus, and the other in the Gum Nebula in Vela; both these can be seen as very faint, flashing points of light. But there can be no doubt at all that a pulsar is the remnant of a formerly very large, luminous star. (Our own Sun will never become a pulsar. It is not massive enough, and will end its career much more sedately by collapsing until it has become a White Dwarf.)

Because pulsars are so faint, and because even their radio signals are not really powerful, they have been hard to track down, and up to 1982 it was impossible to locate any of them outside our Galaxy. Then, however, Dr John Ables, Director of the Parkes Radio Astronomy Observatory in New South Wales, succeeded in identifying a pulsar in the Large Cloud of Magellan. This was very important. It confirmed the very reasonable view that pulsars are not confined to our Galaxy (there is no reason why they should be); and also, the way in which the signals are affected during their long journey of 190,000 light-years tells us a great deal about the conditions in the space between the Cloud and ourselves. Dr Ables' discovery did not 'hit the headlines', but it was nevertheless one of the more significant advances of astronomy during recent years.

92 GALAXIES IN COLLISION?

We live in the Galaxy, sometimes called the Milky Way system. Our Sun is only one of approximately 100,000 million stars in it. It is not the only galaxy; there are many millions of others, of which three are visible with the naked eye.

Two of these, the Clouds of Magellan, lie in the far south of the sky (a fact which European and United States astronomers never cease to regret). They look rather like detached parts of the Milky Way, but they are independent systems well over 150,000 light-years away. Any telescopes will show a great amount of detail in them. The Large Cloud contains one star, S Doradûs, which is at least a million times as luminous as the Sun, though it is so remote that it cannot be seen without optical aid. (It may be the most powerful star known, apart from that celestial freak Eta Carinæ.) The Magellanic Clouds are often regarded as satellites of our Galaxy.

The third naked-eye galaxy is Messier 31, in Andromeda. It is on the fringe of naked-eye visibility, and for some reason Tycho Brahe, last of the great observers of pre-telescopic times, overlooked it; but binoculars show it well. It has been claimed that another galaxy, Messier 33 in Triangulum, can also be seen with the naked eye, though I admit that I have never been able to see it without using binoculars or a telescope.

Below left *Messier 87, the huge elliptical galaxy which is the leader of the Virgo cluster of galaxies.*

Below right *Cluster of galaxies in the southern constellation of Pavo. The distance is about 300 million light years. The cluster includes both spirals and ellipticals.*

Above left *Stephan's Quintet of galaxies (NGC 7317-20)* (Lick Observatory).

Above right *Centaurus A — the nearest radio galaxy.*

Many galaxies are spiral in form. Ours is a well-defined if loose spiral; the Sun, with the Earth and the other planets lies close to the edge of a spiral arm. But there are also many non-spirals. Some systems, such as the Clouds of Magellan, are irregular, but elliptical galaxies are generally very massive, containing more than our own quota of 100,000 million stars but much less star-forming, interstellar material.

Recently some fascinating work has been carried out at the Siding Spring Observatory, in Australia, by David Malin. He has used new and improved photographic techniques on the great 158-in Anglo-Australian Telescope there, and has found that some elliptical galaxies are surrounded by extremely faint 'shells'—so faint, indeed, that they have never been seen or photographed before. It has been found that these shells are made up of stars, and there is a suggestion that shells are formed when two spiral galaxies collide, merging to make one massive non-spiral system and throwing out some of the stars to make the shells.

It is a fascinating idea. It has not been proved, and I doubt whether all elliptical galaxies are the result of collisions between spirals; but there is a great deal about the life-stories of the galaxies that we do not know, and probably a great deal more that we do not even suspect.

93   SPIRIT IN SPACE

Many years ago a great deal was heard about 'empty space'. It was often believed that above the top of the atmosphere there was complete void. In fact this is quite wrong, as was first demonstrated conclusively in 1904 by the German astronomer Hartmann. When looking at the spectrum of Mintaka, the northernmost of the three stars of Orion's Belt, Hartmann found that there were certain lines which

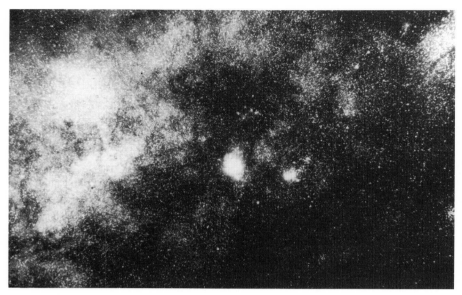

The Milky Way in Sagittarius, showing M.8 and M.20.

behaved in an exceptional way. They did not belong to Mintaka at all; they were due to very tenuous material between the star and ourselves.

As time went by, more sophisticated methods of investigation were developed, and it became possible to identify the substances floating between the stars. Not surprisingly, hydrogen proved to be particularly plentiful, but eventually molecules as well as single atoms were identified, and some of these molecules were found to be organic.

One such molecule is that of ethyl alcohol. Astronomers surveyed an inconspicuous star-cloud in the constellation of Sagittarius (toward the direction of the centre of the Galaxy) and estimated that it contained enough ethyl alcohol to make more whisky than mankind has distilled throughout the history of civilisation.

'Ah,' I can hear some people saying, 'An extra reason for going into space; we can scoop in parts of the interstellar cloud and regale ourselves with whisky throughout the journey!' Unfortunately, nothing could be farther from the truth. The interstellar clouds are incredibly tenuous judged by ordinary standards, and correspond to a density much less than the most perfect vacuum we can produce in our laboratories. This is true even of the 'thicker' clouds: the bright nebulæ, such as Messier 42 in Orion's Sword. If you could take a bucket and plough right through the Orion Nebula, scooping in material steadily, the amount of material collected in the bucket would weigh less than a billiard-ball.

Yet tenuous though they may be, these interstellar clouds are of fundamental importance in astronomy. Visible nebulæ are regions in which fresh stars are being formed, but we must also reckon with the non-visible clouds, which may account for a considerable percentage of the total mass of the universe. It has also been suggested that life on Earth did not begin here, but originated in space and was 'dumped' here by a meteorite or (according to Sir Fred Hoyle) by a comet. Most people believe that this raises more difficulties than it solves, but we cannot rule it out, and the presence of organic molecules between the stars makes it seem less far-fetched than it would otherwise be.

New types of interstellar molecules are being discovered with bewildering rapidity, but I think that the identification of ethyl alcohol is particularly fascinating. Poets and theologians have often eulogised about 'the spirit in space'. Well, the spirit is there—even if in not quite the form that they had hoped!

94 IOSIF SHKLOVSKY AND THE SATELLITES OF MARS

In 1877 Asaph Hall, using the 26-in refractor at the Washington Observatory, discovered the two satellites of Mars, Phobos and Deimos. Both are quite unlike our own massive Moon; they are irregular, cratered bodies, with longest diameters of less than 30 miles. As seen from Mars Deimos, indeed, would look like nothing more than a large, dim star. Phobos is remarkable in as much as its orbital period is only 7 hours 39 minutes—considerably less than a Martian day or 'sol', which is over half an hour longer than ours. Seen from the planet, Phobos would appear to rise in the west, gallop across the sky and set in the east only four-and-a-half hours later.

It was then recalled that Dean Swift, in one of his books about Gulliver's voyages, had described the flying island of Laputa, whose astronomers had said that Mars had two satellites, one of which moved round the planet in less than a Martian day. At that time (1726) no telescope in the world was capable of picking up either Phobos or Deimos. Yet the forecast was not really remarkable. It was known that the Earth had one moon and Jupiter four; so how could Mars manage with less than two?

Then, in 1959, a remarkable theory was published by the Russian astrophysicist Iosif Shklovsky, who had been responsible for fundamental advances in stellar astronomy and whose reputation was second to none. Shklovsky considered the movement of Phobos, which he believed was being slowed down by friction against the upper atmosphere of Mars. Now, Phobos is about 3,700 miles above the surface of the planet (its orbit is almost circular), so that the atmosphere at such a height would be extremely thin. To show indications of 'braking', Phobos would have to be of very low mass.

In his book *Intelligent Life in the Universe*, published together with the American astronomer Carl Sagan, Shklovsky maintained that life was likely to be widespread. So far as Phobos was concerned, he wrote: 'How can a natural satellite have such a low density? The material of which it is made must have a certain amount of rigidity . . . . Thus, only one possibility remains. Can Phobos be indeed rigid on the *outside*, but hollow in the inside? A natural satellite cannot be a hollow object. Therefore we are led to the possibility that Phobos—and possibly Deimos as well—may be artificial satellites of Mars . . . . The idea that the moons of Mars are artificial satellites may seem fantastic, at first glance. In my opinion, however, it merits serious consideration.' After discussing the first Earth artificial satellites, he went on: 'Perhaps we are observing an analogous situation on Mars. According to the distinguished American cosmochemist Harold C. Urey, some

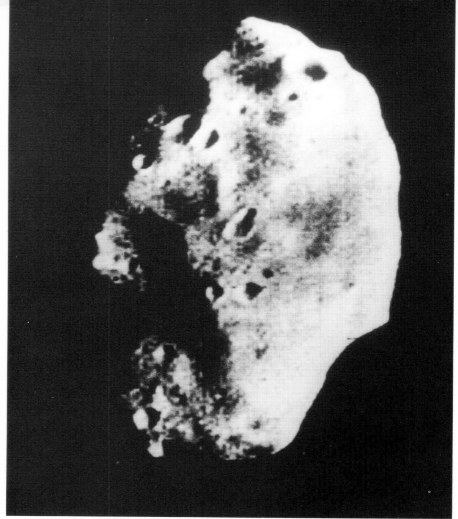

Phobos, the inner and larger satellite of Mars, photographed from close range — and clearly not a space station.

billions of years ago Mars may have possessed extensive oceans suitable for the origin of life. . . . Perhaps Phobos was launched into orbit in the heyday of a technical civilization on Mars, some hundreds of millions of years ago.'

Not surprisingly, this startling theory attracted a good deal of criticism, not all of it favourable! Later, the probes sent to Mars obtained close-range views of Phobos, and showed it to be a rocky, cratered object. Therefore, in 1979 I wrote to Shklovsky, and asked whether he would contribute an article to the *Yearbook of Astronomy*, which I edit. He was kind enough to do so, and began as follows:

'In 1959 I published an article along the lines of my hypothesis about the artificial origins of Phobos and Deimos. It made quite a sensation at the time. It was nothing but a practical joke, however, and the quasi-scientific arguments which I marshalled were of the same type.' He went on to say that since no extra-terrestrial civilisation has contacted the Earth, we are presumably alone in the universe.

This is certainly a complete volte-face. Certainly his statements in the Sagan book, published as recently as 1966, gave no hint that he was indulging in a practical joke. If he was—and we must, of course, accept his word on this—it shows that he has at least a lively sense of humour!

95 TAKE ME TO YOUR (ENGLISH-SPEAKING) LEADER!

From time to time, efforts have been made to pick up artificial signals from beyond the Solar System. Large radio telescopes have been used, and there have been some 'alarms', particularly when the first pulsars were discovered. The regular, fast 'ticking' gave a superficial impression of being non-natural, and the astronomers at Cambridge prudently held the news back until the idea of artificial signals had been disproved. That was the end of the first 'LGM' or Little Green Men theory.

There is every reason to believe that intelligent life is widespread in our Galaxy, and in others. Proof is lacking, and we are quite unable to send probes to other planetary systems and hope to keep in touch with them. If there is to be any direct contact, we must look into methods such as teleportation and thought-travel, which takes us straight into the realms of Dr Who and Lord Darth Vader. Whether this will ever be possible remains to be seen. Meanwhile, there must surely be civilisations much older and wiser than ours. If they have mastered the secret of interstellar travel, and arrive in our vicinity, how will they contact us?

This is where I propose to make a suggestion which may sound outrageous but which is, I maintain, entirely rational. I believe that if aliens decide to visit the Earth, they will call us up in a recognisable language—possibly English—before touching down.

Let me give my reasons. Assume that a piloted probe has come from a planet circling another star, and has entered our Solar System. (Just how the journey is achieved does not matter in our present context.) The first thing that the visitors will do is to find out just what we are like. If they call in during a time of global war, they may decide to retire quietly and hope that we will remain permanently isolated. If not, then they will have to decide how to approach us.

If they are advanced enough to manage a voyage of many light-years, then surely they will have no trouble about listening in to our broadcasts and learning our language. One can visualise the scene aboard the alien ship. Linguists will hold solemn classes, with crew members doing their best to pronounce our words recognisably. Conversation sessions will be held. Examinations may be set. Finally, there will be enough English-speakers to cope with any eventuality, so that actual contact can be made.

Of course there are other problems as well, particularly if the aliens are non-humanoid in appearance; but if they land, blaring out a welcome in straightforward English, we will at least know that we are dealing with beings of advanced type.

This may sound like fantasy, and in AD 1984 it is. But if contact is ever made, the language problem will be a key. An alien who steps out of his space-craft and says 'Wzzzk bdoj oofgj?' is likely to be received with grave suspicion. If he bows politely and says, 'Good morning. I come from Delta Pavonis C. May we have permission to disembark, please?' he will, I hope, be treated with respect. And it could so easily happen—one day.

96   THE MYSTERIOUS NOVA

Among great astronomical observers, Edward Emerson Barnard has an honoured place. He was superbly skilful, and he had the advantage of being able to use very large telescopes, notably the great 36-in refractor at the Lick Observatory. It was with this telescope that he discovered Amalthea, the fifth satellite of Jupiter, on September 9 1892. (Incidentally, this was the last planetary satellite to be found visually; all later discoveries have been photographic, beginning with W.H. Pickering's identification of Phœbe, the outermost satellite of Saturn, in 1898.) But another observation by Barnard remains very much of a mystery.

It was made on August 13 1892—in broad daylight; Barnard was looking at the planet Venus, then in its crescent stage. He wrote as follows:

'I saw a star in the field with the planet. The star was estimated to be of at least the 7th magnitude. The position was so low that it was necessary to stand upon the high railing of a tall observing chair. It was not possible to make any measures, as I had to hold on to the telescope with both hands to keep from falling. The star was estimated to be one minute of arc south of Venus. . . . There seems to be no considerable star near this place, and the object does not agree with any BD star.' (BD stands for the Bonner Dürchmusterung, the best catalogue available at the time.) Barnard gave the right ascension of the object as 6h 50m 21s, declination N.17° 13'.6.

The object was never seen again, by Barnard or anyone else, so that this is the sole record of it. So let us try to decide what it could have been.

We can rule out any satellite of Venus. No satellite exists—and if there had been one, with a magnitude of 7, it could not possibly have been overlooked even before the age of space-probes.

A mistake in the date of the record, then, or perhaps a 'ghost' image of Venus produced in the optics of the telescope? This will not do either. Barnard's records for that night are typically precise, and he was much too experienced and accurate an observer to have been tricked by an optical defect.

We are also able to eliminate all asteroids or minor planets, none of which could have been as bright as magnitude 7 in broad daylight. There seem to be only two possibilities left. The object could have been the nucleus of a comet—but then it would hardly have been so starlike. All in all, it seems that Barnard saw a nova, or new star.

This is plausible. Venus at that time lay in the rich constellation of Gemini, where various novæ have appeared; and at that season of the year Gemini was above the horizon only during the hours of daylight, so that in the ordinary way a nova would have been missed. We may assume that by the time Gemini emerged from the morning twilight, the nova had faded away.

Efforts have been made to check all the stars in the area to see whether any of them shows the characteristics of an ex-nova, but with no success. So the mystery remains, and we are unlikely to solve it now.

97   ARE THERE FISHES IN EUROPA?

Recently I have been noticing a new trend in astronomical thought. It used to be regarded as almost axiomatic that life must be spread widely through the universe,

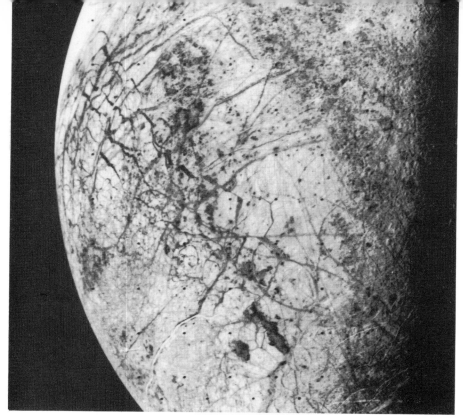

Europa, from Voyager 2, at a range of 150,600 miles. The surface is icy and smooth; the dark features are very shallow indeed.

but this is now being challenged, and I know of at least one eminent astronomer who is convinced that the Earth supports the only life in our Galaxy at least.

If we could prove the existence of living organisms on any single world beyond the Earth, we might regard this particular argument as settled. Unfortunately, this is what we have so far failed to do, and the two formerly most promising candidates—Venus and Mars—have shown no signs of life. But a new entry to the list of possibilities is distinctly surprising: Europa, the second of the four major satellites of Jupiter.

The Voyager probes provided amazing views of all these satellites. Callisto and Ganymede are icy and cratered; Io has a red, sulphury surface and active volcanoes; Europa is very much of an enigma. No mountains; virtually no craters; nothing but an icy surface and strange, shallow 'stripes' which look like mere cracks. It has been said that Europa is as smooth as a billiard-ball, and it is of course very cold. There is no atmosphere; Europa, smaller and less dense than our Moon, cannot possibly retain one.

But though the surface is so unpromising, conditions lower down may be different. And two leading NASA scientists, Steven Squyres and Ray Reynolds, have proposed a startling theory which may sound like science fiction, but which may well be correct—at least in part.

First, let us look at the make-up of Europa. The visible surface is definitely a sheet of ice, probably two or three miles thick. Below the ice there may be an ocean of ordinary water, at least 30 miles deep. The ocean is shielded from 'outer space' by its protective icy covering, but there may be occasional fractures in the ice; water then boils out, freezes, and drops back on to the surface as frost. This, incidentally, would explain why Europa has so smooth a surface. Any irregularities, craters or otherwise, would be covered up.

There is also a good chance that Europa's core is relatively warm. There must be some heat generated by radioactive materials, and also some heat produced by the 'flexing' of Europa by the powerful gravitational pull of Jupiter (which is much more marked with Io, and is presumably responsible for the volcanoes there). Moreover, some solar heat may percolate through when a crack is opened; the fissure would remain for five to ten years before being frozen over again, and Squyres and Reynolds calculate that in each year there should be a total exposed area of at least ten square miles. So to sum up: on this theory Europa has a rocky core, which is surrounded by a deep ocean which is in turn overlaid by a continuous sheet of ice.

Now let us look at our own Antarctic. There are frozen lakes, and in the water underneath there are plenty of organisms. There is enough penetration of sunlight, via fissures, to enable these organisms to survive and multiply. Is it possible that conditions there are roughly similar to those in the hypothetical Europan ocean?

Proof, of course, is lacking. We may be wrong about the make-up of Europa; there may be no subcrustal ocean, and even if there is, it may be totally sterile. Unfortunately there seems little hope of finding out in the foreseeable future. The only probe scheduled to go to the system of Jupiter in the foreseeable future is 'Galileo' (an appropriate name), and even this can do no more than give us closer-range views of Europa's surface. Yet if fissures are observed, and above all if we can glimpse one of the 'geysers' in action, the ocean theory will become very plausible indeed.

We cannot expect fishes inside Europa. But it is not impossible that there is life of a kind; and it will be ironical if we eventually find that while the much-vaunted Mars is sterile, organisms exist in the dim, far-away ocean beneath the ice-fields of Europa.

98 THE GARNET STAR

Of all the naked-eye stars, the reddest is probably Mu Cephei, in the rather obscure constellation of Cepheus, not far from the north celestial pole. Sir William Herschel nicknamed it 'the Garnet Star', and when seen through a telescope it has an appearance which has been likened to a glowing coal.

Like so many red stars, Mu Cephei is variable in brightness. At its best it may rise to magnitude 3.6; at its faintest it goes down to 5.1, though for most of the time it hovers around magnitude 4½. There have been various attempts to assign a period to it, but on the whole it seems to be more or less irregular. Its fluctuations are slow, but are easy to follow, particularly when binoculars are used. A convenient comparison star is its neighbour Nu Cephei, which is white, and is steady at magnitude 4.3.

Mu Cephei is a red supergiant. It is very remote; its distance is rather uncertain, but may be around 1,600 light-years. This means that it must also be very power-ful. Again there is a wide limit of uncertainty, but it is likely that Mu is some 50,000 times more luminous than the Sun.

Its variations are real; the star swells and shrinks, changing its output as it does so. On average the diameter is of the order of 250 million to 300 million miles — perhaps even more. This means that it is even larger and more powerful than Betelgeux in Orion, the most famous of all the red supergiants. But Betelgeux is not nearly so far away, and is 'only' 15,000 times more luminous than the Sun. If Mu Cephei were as close as Betelgeux, it would be splendid indeed.

It is easy enough to find, though only when it is near maximum is its colour really evident with the naked eye. However, any optical aid brings out the red hue well, and it is easy to see why Herschel called it the Garnet Star.

There is, incidentally, one other interesting note. Nu Cephei, which acts as a comparison, is one of the most powerful of all the naked-eye stars. If we accept the distance-estimate of 4,000 light-years, then Nu Cephei is around 100,000 times more luminous than the Sun, in which case it is more powerful than any of the first-magnitude stars with the exception of Canopus. If it were as close to us as, say, Sirius, it would cast shadows. Once again appearances are highly deceptive.

99 STRANGE TELESCOPES

Everyone knows what an ordinary telescope looks like. The refractor, or lens tele-scope, is of the type used for bird-watching; the astronomical refractor is essential-ly the same, though it generally uses a much higher magnification and shows everything upside-down (because the correcting lens, to correct the image, is left out as being unnecessary). A reflector collects its light by means of a mirror, and in the popular Newtonian type the light is sent back up the open tube on to a smaller flat mirror, after which it is directed to the eyepiece set in the side of the tube.

But there have been some strange telescopes. Perhaps the most remarkable of all was built by an Irish nobleman, the third Earl of Rosse, in 1845.

Lord Rosse was an enthusiast, and he had remarkable skill. He aimed to build a reflector with a mirror 72 in in diameter, far larger than anything produced be-fore. But even when the telescope had been made, there was the problem of mounting it; and with the engineering limitations of more than a century ago, Lord Rosse realised that he could not make it 'handy' enough to survey the whole sky. His remedy was to mount it between two massive stone walls, and pivot it at the bottom. That meant that the telescope could see only a narrow strip of the sky. Luckily the Earth rotates, so that the whole of the available sky was brought within range at some time or other—and Lord Rosse used the telescope to make observa-tions of vital importance; he was the first to see that the dim, misty objects which we now know to be galaxies, millions of light-years away, are often spiral in form. The telescope has not been used for over 70 years now, and it has been dismant-led; but the tube and the stone walls are still at the original site at Birr Castle in Central Ireland, and the metal mirror is safe in the South Kensington Science Musuem. One day, with luck, we may get the telescope working again.

Soon after the great Birr reflector was in working order, an attempt to build a really large refractor was made in England. It was due to the Rev John Craig, Vicar of Leamington Spa, who was an enthusiastic amateur astronomer. Craig commissioned a 24-in object-glass from a maker named Thomas Slater, and then had it mounted on a piece of land just south of Wandsworth Prison. It was a weird-looking instrument, with a tube which tapered at both ends; this tube was slung in chains from the side of a solid brick tower, and the eyepiece rested upon a wooden framework which ran on a circular track. Alas, the 24-in lens was so defective that even if it had had a convenient mounting, the telescope would have been more or less useless. Apparently it was dismantled after about six years.

The oddest refractor that I have seen is at the People's Observatory in Treptow, outside Berlin in East Germany. It has a 27½-in object-glass, but the focal length is so long—nearly 70 ft—that the tube has to be slung on struts, with the eyepiece at the pivotal point. It gives every impression of being a huge gun. The mounting, designed by a mechanic named Mayer, weighs 130 tons. The refractor is known officially as the Archenhold Telescope, and was opened on May 1 1896. Candidly, it was never a real success; it was too cumbersome, and I understand that it has not been in use for some years now, but it is certainly an impressive landmark.

Then there was the telescope built in France and shown at the Paris Exhibition of 1901. This was a refractor with a 49-in lens, larger than the biggest object-glass of today (that of the Yerkes 40-in refractor). The Paris telescope was horizontal, so that the light had to be reflected into the lens by a movable mirror. It was a complete failure, and was never used seriously. What happened to the 49-in lens is not known. There is a rumour that it is still stored somewhere in the cellars of the old Paris Observatory; it would be interesting to find it.

Finally, I must mention a 12-in telescope made and used by a friend of mine. He lives in a house where there are inconvenient trees to all sides, and to make his telescope fully mobile he has mounted it upon a wheelbarrow. I am delighted to say that it works excellently!

100 VREDEFORT RING AND THE ARAL SEA: IMPACT CRATERS?

In 1979, during a visit to South Africa, I flew in a helicopter over the Vredefort Ring. This is a large structure to the north of Pretoria, and it gives the impression of being a very worn-away crater—an impression which is even stronger when you look at it from the air. Inside it there are two villages, Parys and Vredefort itself.

What caused it? There have been many suggestions that it is an impact crater, blasted out long ago by the fall of a giant meteorite. The famous crater in Arizona, nearly a mile wide, is the classic example of a meteorite crater. Wolf Creek, in Australia, is another crater which is undoubtedly of impact origin.

Incidentally, the largest known meteorite is still lying where it fell, in prehistoric times, in Hoba West in Southern Africa, near the town of Grootfontein. It is hardly likely to be moved, as it weighs over 60 tons. But is the Vredefort Ring an impact crater?

Geologists who have carried out long-continued studies of it are almost unanimous in saying that it is not. It gives every indication of being a volcanic structure. The same is true of many other alleged impact craters, and we have to be very cautious.

In 1981 a Russian scientist, Dr Borizov, put forward the intriguing idea that the Sea of Aral is an old impact crater. He claims that the whole of the Aral Basin shows indications of this, and that wells are found to contain meteoritic material. If so, then the crater's age must be at least 40 million years, and the meteorite which made the crater would have weighed many thousands of tons.

I have not been to the Sea of Aral, so that I can give no personal opinion. For that matter, arguments about the origin of the craters of the Moon still continue. No doubt some of them are due to impacts and others to vulcanism, but even the Apollo missions have not given a decisive answer.

There is always the chance of a major new meteoritic impact, and it could be a serious affair, but fortunately the chances of the Earth being hit by a really large missile in the foreseeable future are so slight that they need give rise to no alarm. Space is a large place, and the Earth is a relatively small target.

101 FRAGMENTS FROM MARS?

I wonder if any of you have been to Antarctica? I haven't. Frankly I would like to go, but it is not exactly hospitable or easily accessible. The whole continent is covered with an ice-sheet, and until the fairly recent arrival of scientists the local penguins were left in peace. Now, it has been found that Antarctica is a veritable treasure-house of meteorites.

Meteorites are solid bodies which plunge through the Earth's atmosphere and land without being burned away. They are quite different from meteors, which destroy themselves in the luminous streaks which we call shooting-stars; the average meteor is smaller than a sand-grain, while the largest known meteorite weighs over 60 tons. Moreover, meteorites are of different types. Some are stony; others are rich in iron. The iron meteorites are the more durable, and survive unharmed, while stony meteorites are often eroded away after landing to a state in which they cannot be recognised.

Many meteorites have been found, but Antarctica is a particularly good place to look, because the meteorites coming down there are preserved in ice. Since 1977 there have been several very successful meteorite hunts.

Fragment of the Barwell Meteorite found by me at Barwell village.

The Leonids, from Arizona, March 17 1966. The meteors appear as streaks because this was a time-exposure.

It is generally believed that meteorites represent material left over, so to speak, when the Earth and the other planets were formed out of a cloud of dust and gas associated with the youthful Sun, over 4,500 million years ago. However, there have also been suggestions that some meteorites may have come from the Moon—either hurled out from lunar volcanoes, or else blasted off the lunar surface by the impacts of particularly large missiles. Nobody is at all certain (frankly, I admit to being sceptical), but now there has been an even more intriguing idea put forward. Nine of the meteorites found in Antarctica are suspected of having come not from the Moon, but from Mars. They appear to be much younger than most meteorites, and date back only about 1,300 million years instead of the more usual 4,500 million years. This seems to rule out a lunar origin, because the Moon's active period ended well before 2,000 million years ago. But Mars does have huge volcanoes—one of them, Olympus Mons, rises to three times the height of our Everest—and since these nine peculiar meteorites are thought to be rather similar in composition to the Martian surface, a Martian origin has been proposed.

I wonder! It would need a tremendous outburst, and whether the Martian volcanoes were ever capable of such force we do not know. Again, I must confess to being extremely doubtful. But I may be wrong, and it is fascinating to think that there is just a chance that the meteorite hunters of Antarctica are collecting fragments from Mars.

102   ALBIREO

It is an interesting fact that a surprisingly large number of stars are either double or multiple. It has even been suggested that single stars such as our Sun are in the minority. Some of the doubles are strikingly beautiful; and in my opinion the loveliest of all is Albireo, known officially as Beta Cygni.

Albireo is easy to locate, because it is one of the five stars making up the Cross of Cygnus, the Swan. The brightest member of the Cross is Deneb, which is of the first magnitude and is, incidentally, a real cosmic searchlight at least 60,000 times as luminous as the Sun. The Cross is quite evident; in fact Albireo rather spoils the symmetry, because it is the faintest of the five and is further away from the centre of the pattern. It lies rather above an imaginary line joining the two brilliant stars Vega in Lyra (the Lyre) and Aquila (the Eagle), and Cygnus is well to the north of the celestial equator, so that it is well seen for a large part of the year. In fact, Deneb never actually sets as seen from the latitude of London.

With the naked eye there is nothing remarkable about Albireo. But if you have a telescope, you will see at once that it is made up of a golden-yellow star with a vivid blue-coloured companion. Even binoculars will separate the pair; the apparent distance between them is 35 seconds of arc, which, for a double star, is very wide.

Both components are powerful. The yellow primary is equal to 700 Suns put together, and even the fainter blue companion shines as powerfully as 100 Suns. Clearly, then, they must be very far away; their distance from us is almost 400 light-years, so that we are now seeing them as they used to be when England was ruled by Queen Elizabeth I. They really do make up a connected or binary system but the real distance between them is about 400,000 million miles. And there is a suspicion that the bright yellow star may itself be a very close double.

Imagine the scene from a planet orbiting either of the Albireo pair! A brilliant yellow Sun, with a glorious blue one; the shadow effects would be magnificent. But it would also make for a rather uncomfortable and variable climate, so that on the whole it is just as well for us that our Sun is a sedate single star.

103   THOSE PLAQUES: HELPFUL OR MENACING?

At the moment there are four probes, two Pioneers and two Voyagers, which are on their way out of the Solar System. Each carries a plaque which is designed to tell any alien civilisation just where the space-craft came from; the Voyagers carry records as well. And, strangely, there have been suggestions that this procedure is dangerous. One astronomer has even recommended that the probes should be destroyed, if possible, before they can tell any alien races where we are to be found.

Let me say at once that I regard these ideas as not only sensational, but frankly ridiculous. For one thing, none of the probes can encounter another star for

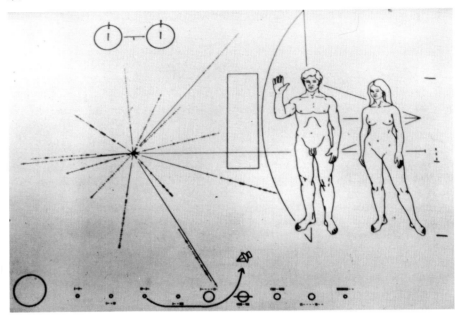

The Pioneer Plaque, now on its way out of the Solar System.

thousands of years, and even then the chances of their being located and taken in for inspection are, to put it mildly, slim. But in any case, what are the so-called dangers?

The idea seems to be that if an alien race finds out where we are, they may descend upon us, Stars Wars fashion, and take us over, so that we could end up at the mercy of a being no more benevolent than Lord Darth Vader.

To me, this is sheer nonsense. We do not have any definite proof that other civilisations exist. I believe they do; the Earth is an ordinary planet moving round an ordinary star, and there are 100,000 million stars in our Galaxy alone, so that to regard ourselves as unique seems both conceited and illogical. On the other hand it will be very difficult to obtain definite evidence. Certainly we cannot hope to establish contact by using rockets; it would take far too long, and we must have recourse to some method which is so totally beyond our powers at present that we cannot even speculate about it.

If we ever manage to achieve such a thing, we will have learned enough to have put the futility of war far behind us. (If we do not succeed in doing this within the next few centuries, we cannot hope to survive.) The same is true for any alien race. In fact, any remote civilisation advanced enough to recover one of our space-craft, identify its planet of origin and then pay us a visit, will inevitably be so far in advance of ourselves that it will certainly come in a spirit of friendship rather than conquest.

So the Pioneers and Voyagers are no threat. Whether they will ever be found seems highly dubious, but in any case we need not be apprehensive. In fact, I would say that any race capable of contacting us would be able to teach us a great deal!

104 PLUTO/CHARON: A DOUBLE ASTEROID?

Pluto is usually listed as 'the outermost planet'. Certainly its mean distance from the Sun is much greater than that of Neptune, the most remote of the giants; it takes 248 years to go once round the Sun, as against only 164¾ years for Neptune. But Pluto has a strange orbit, much less circular than those of the other planets. Between 1979 and 1999 it moves closer in than Neptune, and will pass perihelion in 1989. For the moment, therefore, it does not mark the boundary of the known planetary system.

In 1977 James Christy, of the United States Naval Observatory, discovered that Pluto's photographic image is not regular in outline; he saw that it was elongated in a way which made him suspect that he was looking at a double object. In fact, Pluto had a satellite. It was even given a name: Charon, in honour of the gloomy ferryman who used to take new arrivals to King Pluto's underworld realm across the river Styx. Charon seemed to be at least one-third the diameter of Pluto itself, and to orbit it at a distance of less than 12,000 miles. But did Charon really exist as a separate body, or could Pluto be genuinely oval in form?

For some time the problem remained unsolved, but confirmation of Charon came in 1981. First, on April 6 of that year Pluto almost passed in front of a star, hiding or occulting it. I say 'almost' because the occultation by Pluto did not take place. I watched the phenomenon from my observatory at Selsey in Sussex, using my 15-in reflector, and watched Pluto as it tracked past the star. But at the South African Astronomical Observatory in Cape Province, Alastair Walker saw a brief occultation; the star 'winked' for 50 seconds. Pluto was not in the right position to occult the star, but Charon was. Therefore, Charon existed.

The system of Pluto and Charon to scale.

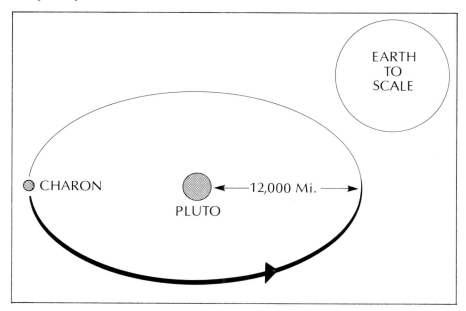

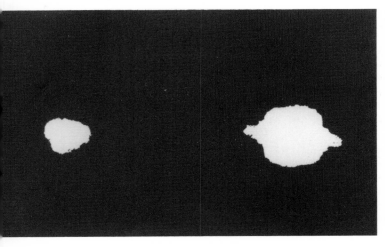

Pluto and Charon, shown separately in this electronic view obtained from the Mauna Kea Observatory in Hawaii. The 'spokes' from Pluto are instrumental effects, not rings.

Final proof was obtained by two French astronomers, D. Bonneau and R. Foy, working with the large CFH or Canada-France-Hawaii telescope on the top of Mauna Kea. They photographed Pluto using what is called the speckle interferometry technique. This involves taking a great many short-exposure pictures, enhancing them electronically, and then 'averaging them out'. They were successful in showing Pluto and Charon definitely separated.

Though the existence of Charon is no longer in doubt, it and Pluto are very small by planetary standards. Pluto cannot be more than 1,800 miles in diameter (smaller than the Moon), and Charon no more than 700. Quite possibly the pair should rank as a double asteroid rather than a double planet, particularly as Charon goes round Pluto in the same time that Pluto takes to spin on its axis (6 days 9 hours), so that as seen from Pluto, Charon would appear fixed in the sky. These remote worlds are unquestionably strange, but they are of tremendous importance in our efforts to understand the mysteries of the Solar System.

105 ZETA PHŒNICIS

To my mind, one of the most confusing regions in the entire sky is that of the Southern Birds. There are four bird-constellations: Grus (the Crane), Phœnix (the Phœnix), Pavo (the Peacock) and Tucana (the Toucan). All are invisible from England, so that to see them you have to travel south. Grus is easily the most distinctive of the four. It lies not far from the bright, rather isolated Fomalhaut in the Southern Fish (which does rise briefly over the London horizon), and contains a curved line of stars, giving a vague impression of a bird in flight. Of its two main stars, one (Alnair) is white and the other (Al Dhanab) warm orange.

Phœnix is much less distinctive. It has one second-magnitude star, Alpha Phœnicis or Ankaa, and nothing else above magnitude 3. However, one of its stars, not dignified by a proper name and known merely as Zeta Phœnicis, is of considerable interest. It lies fairly near the brilliant star Achernar in Eridanus (the River), which is the best way to locate it.

Normally, Zeta Phœnicis is of magnitude 3½. (I am assuming that everyone knows what is meant by 'magnitude': it is a measure of a star's apparent brightness. 1 is brighter than 2, 2 brighter than 3, and so on. Below magnitude 6 you need binoculars or a telescope.) Every 1 day 16 hours, Zeta Phœnicis fades. It drops down to below magnitude 4, so that with any appreciable moonlight around you will be hard pressed to see it with the naked eye. It does not stay faint; it soon slowly brightens up again, after which nothing more happens for another 1 day 16 hours.

Zeta Phœnicis fluctuates in brilliancy, but it is not a true variable star. It is made up of two stars, one brighter than the other, and so close together that they cannot be seen separately even with a powerful telescope. When the fainter member of the pair passes in front of the brighter, the total amount of light which we receive falls—and this explains the regular fading.

Zeta Phœnicis is not unique. There are many more of these 'eclipsing binaries' in the sky, the most famous being Algol in Perseus. Zeta Phœnicis is not exceptional; it is rather over 200 light-years away, and the total luminosity of the system is about a hundred times that of the Sun.

All the same, Zeta Phœnicis is worth looking at. If you know when a fade is due, you will be able to check it by observing every ten minutes or so and checking the magnitude against the neighbouring stars which do not vary. Certainly its behaviour would seem strange if the reason for it were not known. No wonder that Algol, the first-discovered member of the class, has often been nicknamed the Winking Demon.

106   THE SURFACE OF TRITON

Triton, the large satellite of Neptune, is a peculiar world. Because it is so far away, and because no rocket probe has yet been anywhere near it, our knowledge is very limited, and we are not even sure of its size. There seems little doubt that it is larger and more massive than our Moon, but estimates of its diameter range between 2,200 miles and as much as 3,600 miles. Probably the smaller value is the more likely.

The first unusual fact about Triton is that it moves round Neptune in a retrograde direction: that is to say, opposite to the direction in which Neptune itself spins. Secondly, there have been suggestions that its lifespan is limited, and that between ten million and 100 million years from now it will have come so close-in that it will be broken up by the powerful pull of Neptune's gravity.

Meanwhile, what is Triton like? It appears only as a tiny disk in our telescopes (which is why its diameter is so difficult to measure), but some results can be obtained by using the spectroscope. This has been done with Pluto, which is probably rather smaller than Triton and is covered with a layer of frost—not ordinary frost, but methane frost. Triton, too, must have some methane frost, but not so much as with Pluto, and it may be confined to one hemisphere only. Triton takes rather less than six days to complete one journey round Neptune; this is also the length of its own 'day', because, like all major satellites, it keeps the same face turned permanently toward its planet.

A real surprise has been the discovery of nitrogen on Triton. The evidence indicates that this nitrogen is not in the form of a gas (remember that Titan, the senior satellite of Saturn, has a dense, nitrogen-rich atmosphere), but is liquid. There may even be a very shallow nitrogen sea over at least part of Triton's surface. It is not likely to be more than a few feet deep, and beneath it the surface may well be rocky.

Yet even if all this is correct, the situation is bound to change in the future if Triton spirals closer and closer to Neptune. Well before it is finally broken up all its atmosphere (if it has one!) would be stripped away, and so would the methane sea. Of course, this will not happen for a long time yet, and recent investigations indicate that it may not happen at all.

107 SWITCH-OFF ON THE MOON

As most people will know, the first men to reach the Moon did so in July 1969. Neil Armstrong, commander of the Apollo 11 mission, stepped out onto the grey plain of the Mare Tranquillitatis or Sea of Tranquillity, closely followed by Colonel Edwin Aldrin. Since then there have been five further successful landings. Each expedition set up recording equipment on the lunar surface. An ALSEP (Apollo Lunar Surface Experimental Package) contained instruments of all kinds, ranging from solar wind detectors to 'moonquake recorders' or seismometers. Most of these instruments continued working for a long time after the end of the Apollo programme, and the results were received back on Earth.

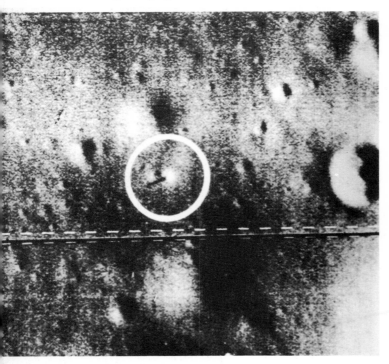

Left *Surveyor 1 on the Moon's surface, photographed from Orbiter 2 on February 22 1967. Surveyor is the white object casting a shadow, which is 30 ft long. Surveyor 1 is still there, and will remain until somebody collects it.*

Above right *The instruments left on the Moon from Apollo 14, near Fra Mauro. The astronauts were Alan Shepard and Thomas Mitchell. Part of their hand-pulled 'cart' is visible on the left.*

Right *Edwin Aldrin deploying the lunar seismic experiment from Apollo 11, photographed by Neil Armstrong.*

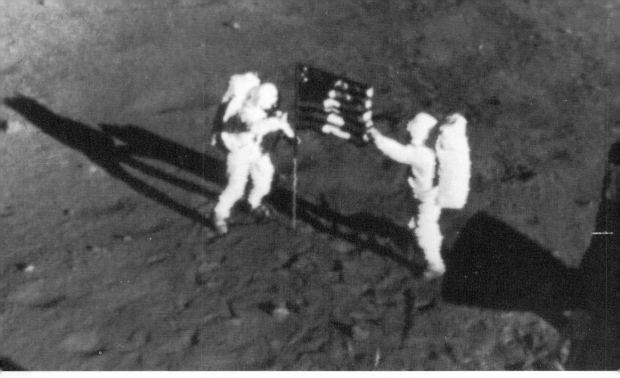

Neil Armstrong and Edwin Aldrin plant the American flag on the surface of the Moon.

I was particularly interested in the moonquake records, because for many years I had been one of a team studying the small, elusive glows and obscurations in various areas of the Moon which I originalled Transient Lunar Phenomena or TLP—a term which has now come into general use. We thought that there must be an association between the TLP areas and the moonquake sites, and we were right. The records agreed well, which was highly satisfactory.

It was too good to last. In 1977, to the intense disappointment of lunar enthusiasts, the recording equipment on the Moon was deliberately switched off. Some of the instruments were showing signs of failure (not surprising, after several years), but others were continuing to work well, and the sudden cut-off was most frustrating. The main reason was, of course, money. The funds of NASA, the American space agency, had been drastically reduced, and economies had to be made. So the Moon is now silent once more; we are no longer receiving signals from it.

It must be admitted that the Apollo programme had done everything that it was capable of doing. In a way the decision to discontinue it after number 17, instead of going on as far as Apollo 21 as had originally been planned, was logical, mainly because there was no rescue provision, and the fate of an astronaut stranded on the Moon does not bear thinking about. (There had already been one mishap, with Apollo 13, when an explosion in the main service module wrecked the principal engines. Had the explosion happened on the return journey the result would have been disaster, because the astronauts would no longer have had the Lunar Module whose engine was eventually used to bring the crippled space-ship home.) So for the moment we must wait in patience. No doubt men will return to the Moon, and a permanent base will be set up there, but this lies in the future.

108 RED STARS ?

I have already described the curious case of Sirius, which was described as a red star by ancient astronomers even though it is now pure white (see article 53). Is this the only case?

During the last century G.F. Chambers, a London barrister and amateur astronomer, published a number of books. They are extremely good, and contain information not easily obtainable in condensed form elsewhere, so that even though they are so out of date they still retain their value. In the 1890 edition, Chambers published a list of bright stars which appeared definitely red. It therefore occurred to me that it would be worth checking up to see if there were any more strange discrepancies; and I found several.

The catalogue contains 62 stars of magnitude 4 or brighter (I did not consider stars fainter than that, because colours of stars below magnitude 4 are almost impossible to detect with the naked eye). In most cases the spectra of Chambers' 'red stars' were of types K or M. Only 12 of them were of 'earlier' spectral type, and so would not be expected to show obvious redness. These twelve were: Lambda Andromedæ (G8), Eta Aræ (B8), Upsilon Boötis (G5), Delta Cephei (F8), Eta Canis Majoris (B5), Delta Crateris (G8), Beta Doradûs (F-G), Achernar or Alpha Eridani (B5), Epsilon Leonis (G0), Gamma Hydræ (G5), Omicron Tauri (G8) and Mu Velorum (G5).

Of course, stars of type G (of which our Sun is one) are regarded as yellowish; they could hardly be classified as 'red'. Incidentally, Chambers' catalogue also includes Alpha Centauri, the nearest bright star beyond the Sun, describing it as 'red or rich yellow'. It is a binary; the spectral types of the two components are K1 and G2, and the pair gives no impression of redness to me, either with the naked eye or with binoculars or a telescope.

Even more remarkably, however, the catalogue includes Achernar, Eta Canis Majoris and Eta Aræ, which have B-type spectra; their surfaces ar very hot indeed, and they are pure white stars. Nobody looking at Achernar, in particular, could describe it as having any colour at all, and the same is true of the highly luminous Aludra or Eta Canis Majoris (over 50,000 times as powerful as the Sun), even though from England it is always rather low down. Achernar and Eta Aræ, of course, are too far south to rise from anywhere in Europe.

I think the inclusion of Achernar may be significant. Chambers was a careful writer, but presumably he had never been far south enough to see Achernar, so that he had to take his description second-hand. It is in this way that mistakes can arise, and can even be perpetuated. Achernar has shown no colour during the history of mankind, and if this sort of error can appear in a reputable book it indicates that other estimates of colour given in past lists must be treated with considerable reserve.

109 THE COMET THAT NEVER WAS

On May 14 1784 Charles Messier, the famous comet-hunter, received a letter from the Chevalier Jean August D'Angos, Knight of Malta, reporting the discovery of a faint comet in the constellation of Vulpecula, the Fox. D'Angos gave two positions for it, and then sent Messier a list of its orbital elements. Messier looked for the comet, but could not find it. This is hardly surprising, as the comet never existed!

D'Angos was born at Tarbes, in the Pyrenees, in 1744. He served as a captain in the pre-Revolution French Army, and became a Knight of Malta. He was also elected an associate of the French Academy of Sciences, and had something of a reputation as an astronomer and a chemist. In fact his chemical experiments were disastrous. The Grand Master of the Knights of Malta invited him to set up an observatory and laboratory in the Palace of Valetta. D'Angos did so, and one of his experiments involving phosphorus resulted in a most unsatisfying explosion which wrecked not only the observatory, but much of the Palace as well. D'Angos returned to France, where he remained until his death at Tarbes in 1833.

What, then, about his comet? In 1806 the Paris astronomer J. Burckhardt attempted to re-calculate its orbit. D'Angos was asked to send his observations; he replied that all of them had been lost. When Burckhardt made his calculations from the only observations available (those which D'Angos had sent to Messier) he found that the orbit was quite unlike that given by D'Angos.

The next development came with the discovery of a note by D'Angos, published in a German periodical, giving no less than 14 positions for the comet! The problem was taken up by Johann Encke, a leading German astronomer, who found that by using the orbital elements given by D'Angos he could reproduce all 14 positions—but only if he used a value for the length of the astronomical unit (that is to say, the distance between the Earth and the Sun) which was exactly ten times too large. In fact, all of D'Angos' observations had been faked. Encke summed up matters up by saying that 'D'Angos had the audacity to forge observations which he had never made of a comet which he had never seen, based upon an orbit which he had deliberately invented, all to give himself the glory of having discovered a comet.'

In 1798 D'Angos wrote to several French astronomers, claiming that on January 18 he had seen a comet pass across the face of the Sun. He added that he had seen a similar transit in 1784, and that both observations must refer to a comet which had been seen in 1672 and had been thought to have a period of 14 years. Unfortunately for him, he based this conclusion upon an orbit for the 1672 comet which had been published years earlier but which had contained a misprint. D'Angos, predictably, used the faulty orbit. . . .

This was, apparently, his last foray into cometary astronomy. Fortunately, his career seems to be unique.

110   LOST AND FOUND

The minor planets or asteroids, most of which keep to the region of the Solar System between the orbits of Mars and Jupiter, have not always been popular with astronomers. Photographic plates exposed for quite different reasons are often found to be crawling with asteroid trails, and one irritated German even went so far as to describe them as 'vermin of the skies'. This seems unfair, and today the importance of the asteroids is fully appreciated. More than 2,600 have had their orbits worked out, and have been given official numbers.

It is clearly not an easy matter to keep track of them all, and an asteroid is numbered only when it has been observed well enough for it to be recovered in future years. There are in fact only six numbered asteroids which have been 'lost': they are 473 Nolli, 719 Albert, 724 Hapag, 878 Mildred, 1026 Ingrid and 1179 Mally. The loss of Albert was a pity, because it was one of those asteroids which swings well inside the orbit of Mars and makes close approaches to Earth. Unfortunately it was seen at only one approach, that of 1911, and it is so small that its recovery now will be largely a matter of luck.

However, there have been some remarkable successes. Thus Number 1537, Transylvania, 'went missing' between 1940 and 1981, and the much fainter 452 Hamiltonia was not seen from 1900 until it was picked up once more in 1973. Number 330, Adalberta, never existed at all. It was 'discovered' by Max Wolf in 1892, but was photographed only twice, and later investigations showed that the images were of two separate stars rather than a single moving asteroid. Of the unnumbered asteroids, the main loss is Hermes, which brushed past us in 1937 at a mere 485,000 miles—less than twice the distance of the Moon.

Among other 'Earth-grazers' there is Apollo, which has given its name to a whole class of asteroids. It was discovered in 1932, but was followed until only May 15 of that year, when its distance from the Earth was seven million miles. That was the last seen of it until 1973, but its possible position had been worked out, and it was duly recovered by R. McCrosky and Cheng-Yuan Shao, using the 61-in reflector at Harvard. It had been missing for 41 years, and as it is no more than a mile in diameter it is always faint. At its last return, in 1981, it was actually contacted by radar, and there is little fear of it being lost again. Even smaller is Adonis, half a mile across, which was discovered in 1936 and was found again by Charles Kowal, at Palomar, on February 14 1977, when it was 13 million miles from us.

One asteroid which I hope will not be mislaid is number 2602, discovered by Dr E. Bowell on January 24 1982 and which was named 'Moore' after me! It belongs to the main swarm, and has a diameter of perhaps three miles. I have yet to see it, as it is beyond the range of my telescope, but one day I hope to make the acquaintance of what I am bound to regard as my own personal part of the Solar System.

111 COMETS WITH DUSTY TAILS

Many short-period comets are known. They move round the Sun in orbits which are, in most cases, very elliptical; they have periods of a few years or a few tens of years, and we always know when and where to expect them. With one exception (Halley's Comet) they are faint, and generally look like tiny patches of haze in the sky even when viewed with large telescopes.

One of these is Comet Tempel 2, which has a period of 5.3 years. It was discovered in 1873 by the German astronomer Ernst Tempel, and has so far made 17 appearances, the last of which took place in 1983. No tail had ever been detected. Then, in the summer of 1983, it was examined by the equipment on the Infra-Red Astronomical Satellite, IRAS. The results were surprising. Tempel 2 does have a tail, made up of what we may call 'dust' but it is too faint to be seen visually, and is detectable only in infra-red. The length of the tail is about 20

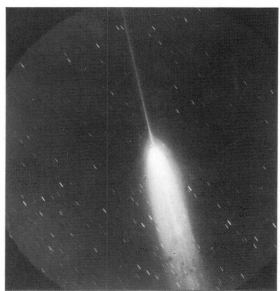

Above *Morehouse's Comet, photographed by E.E. Barnard on November 16 1908. This was a very active comet, showing rapid changes in its tail.*

Left *Comet Arend-Roland, 1957, photographed by E.M. Lindsay. This was a naked-eye comet. Note the 'reverse tail' — not actually a tail at all, but debris in the comet's orbit catching the sunlight.*

million miles, not far short of the minimum distance between the Earth and the planet Venus.

This discovery was unexpected. Comets, of course, are ghost-like things; they are made up of 'ices' together with small particles and very tenuous gas, and most of their mass is concentrated in the nucleus, which is no more than a few miles in diameter. Tails are formed when the ices in the comet's head begin to evaporate as the comet nears the Sun, and material streams out (always in the direction opposite to the Sun, so that when a comet is moving outward it actually travels tail-first). It had not been thought that faint periodical comets such as Tempel 2 had enough material left to produce tails.

161

Probably not all faint comets have similar tails. The evidence is against it, because another periodical comet, Tempel 1 (discovered by Ernst Tempel in 1867) was also on view, and IRAS detected no tail. On the other hand, another very dim comet actually discovered by IRAS showed a considerable dust-tail. This particular comet was discovered at about the same time by two ground-based astronomers, George Alcock in England and Araki in Japan, so that its official designation is IRAS-Araki-Alcock. It was small, but came close to the Earth, and was even visible with the naked eye for a few nights. It has a period of many centuries, and so we will not see it again.

At the moment we do not really know whether dust-tails are common or not, but at least the IRAS results show us that even comets are likely to provide us with plenty of surprises.

112 NEPTUNE: SATELLITE NUMBER THREE?

Neptune is the outermost of the giant planets. (Indeed, at the moment it is the outermost of all the planets, because between 1979 and 1999 Pluto is actually closer-in to the Sun.) Large though Neptune is, it is too faint to be seen with the naked eye. Good binoculars will show it easily as a starlike point, but not even large telescopes will show anything definite on its pale bluish disk. Its real diameter is around 30,000 miles—much larger than the Earth, slightly smaller than its non-identical twin planet Uranus.

Neptune is a gas-giant. It may have a rocky core, but it contains a great deal of methane and water as well as an atmosphere rich in hydrogen. Certainly nothing could land there; any hopes of visiting Neptune are doomed to disappointment.

There are two known satellites. One, Triton, is a big world, larger than our Moon. It goes round Neptune in an almost circular path, but it travels in a wrong-way or retrograde direction, like a car going the wrong way round a roundabout. The other satellite, Nereid, is much smaller, and has an elliptical orbit more like that of a comet than a satellite.

Most of the outer planets have extensive satellite families. Jupiter has 16, Saturn at least 20 and Uranus five, so that Neptune seems rather ill-served with only two! But a third has recently been suspected, in a rather indirect way.

In May 1982 Neptune passed in front of a star, hiding or occulting it briefly. The event was observed by two teams of astronomers from the University of Arizona. In addition to the main occultation, the star was briefly blotted out by what may have been an unknown body—in fact, a third Neptunian satellite. If this is correct, the satellite will be around 60 miles in diameter, and will be moving in an orbit about three times the diameter of Neptune itself.

There is nothing really unexpected about this, but at the moment there is no confirmation, and if it exists the satellite will be very faint. Probably our best chance of confirming it lies with Voyager 2, the unmanned rocket probe which is now wending its way into the outer reaches of the Solar System. It is due to encounter Uranus in 1986, and then Neptune in 1989. If it continues to work well, we may have a great deal of new information about Neptune and its system within the next decade. If not, then I am afraid we may have to wait for a long time before deciding whether satellite number three really exists or not.

113   THE HUNT FOR PLANET X

It is now more than half a century since Clyde Tombaugh discovered Pluto. He was working at the Lowell Observatory in Arizona, using a 13-in telescope installed specially for the search, and he attacked the problem systematically. Pluto turned up not very far from the position given by Percival Lowell, who had died in 1916 and who had worked out the location of his 'Planet X' mathematically, much as Adams and Le Verrier had done for Neptune so long before. Actually, Tombaugh had not selected this particular region because it had been indicated by Lowell; he was working painstakingly along the Zodiac. This discovery was due entirely to his patient, thorough searching. Had he been even marginally less on the alert, Pluto would have slipped through his net.

But problems soon arose. Pluto seemed to be surprisingly small and light-weight. The discovery of its satellite, Charon, in 1977 gave final proof that it is not nearly massive enough to cause measurable perturbations in the movements of giant planets such as Uranus or Neptune. In fact, Pluto is not Lowell's Planet X. Either his reasonably correct prediction was sheer luck, or else the real Planet X remains to be found. (Note, by the way, that a similar result had been reached by W.H. Pickering, an energetic planetary observer. A search was made at Mount Wilson Observatory on the basis of Pickering's calculations, and Pluto was subsequently found in the plates; it had been missed because of the two recorded images, one of which fell upon a star and the other upon a flaw in the plate.)

Frankly I am sceptical about the 'luck' theory; it would be too much of a coincidence. This means that Planet X really exists. Clyde Tombaugh continued his

Pluto, photographed by Clyde Tombaugh in 1930. The planet is arrowed, and has clearly moved between the two exposures, which were not taken on the same night.

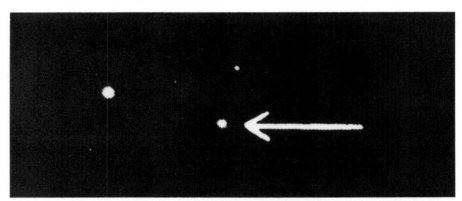

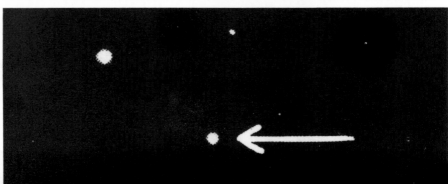

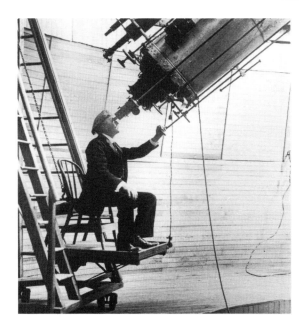

Percival Lowell at the 24-in refractor at Lowell Observatory, Flagstaff, Arizona.

hunt long after the identification of Pluto; altogether he examined the images of 90 million stars (!) and if there had been another trans-Neptunian planet above magnitude 17 he would certainly have found it. So Planet X must be fainter than that.

To carry out a thorough search, going down to—say—magnitude 21 would be a truly Herculean task, and we have very little chance of locating Planet X unless we have at least a vague idea of where it may be. This is where I am bold enough to make a suggestion, though with the full knowledge that it is wildly speculative. It depends upon dismissing the 'luck' theory (otherwise it fails at once). Suppose that in 1930, when Pluto was found, it and Planet X were approximately lined up, so that Planet X was in the same region of the sky but was below the limit of detection? This gives us at least some kind of starting-point.

Now consider Bode's Law, which has caused such tremendous argument. The Law is a mathematical relationship between the distances of the various planets from the Sun, but it breaks down completely for Neptune, the third most massive planet in the Solar System, and frankly I have not the slightest faith in it. If Planet X exists, and is of about the same size as Uranus or Neptune (around 30,000 miles in diameter), it could well be about as far beyond Neptune as Neptune itself is from Uranus. Starting from the assumed 1930 position, we can then estimate how far it will have moved by now. If it is not too far from the ecliptic, we reach a position in the constellation of Leo, not far from the faint star Chi Leonis.

Of course, this is really little more than guesswork, and is hopelessly inadequate to persuade anyone to use powerful equipment in what would probably be a vain search. But there really does seem to be something unexplained in the movements of Uranus and Neptune. All I will say is that if Planet X is ever found, and really was near Chi Leonis in the early 1980s, I will give myself a modest pat on the back!

114   A VERY YOUNG STAR

When a star first condenses out of interstellar material, usually in a gaseous nebula, it is unstable; it varies irregularly, and is often surrounded by a shell or cocoon of dust. Before it settles down to sober middle age, it blows this shell away. T Tauri, in the constellation of the Bull, is such a star, and has given its name to the whole class.

There are many T Tauri stars inside the great Orion nebula, which is nothing more nor less than a stellar nursery. One variable, FU Orionis, brightened up abruptly in 1939 from magnitude 16 to magnitude 10, and has not altered much since then. Just as remarkable is a star in another nebula—the so-called North America in Cygnus, the Swan, not far from Deneb in the sky. The star is known as V.1057 Cygni, because it is the 1,057th variable star to be catalogued in Cygnus. Before 1969 it was very faint, and fluctuated between magnitudes 15½ and 16½, so that it was beyond the range of most telescopes of the size used by amateurs. Then, however, it increased to the 10th magnitude. It was not actually 'caught in the act', but its behaviour was recorded on photographs which were examined slightly later; the rise took about 250 days.

This was not simply caused by the blowing-away of a dusty shell. The spectrum changed from type K to type A, and probably the star itself swelled out. Detailed photographs show that it is now attached to a small, bright patch of nebulosity that was not visible before the rise in brightness, and seems to represent dust near the star which has only now been lit up dusty material of this sort emits infra-red radiation, and can be detected.

My interest in V.1057 goes back to 1971. An astronomer friend of mine, Dr Martin Cohen, was working in the United States and was observing the star in infra-red; he asked me to keep a watch on it and see how it behaved in ordinary light. I have followed it carefully ever since, and the results have been somewhat unexpected. When I began my observations I estimated the magnitude as 10.7, but from mid-1973 it started to decline, and continued to do so until by the beginning of 1984 I gave its magnitude as below 12.

What is happening to it? Will it go on fading until it has returned to the pre-1969 level—and if so, why? Time will tell; but at all events V.1057 Cygni is most unusual, even for a star which is so youthful by cosmic standards.

115   THE CENTRE OF THE GALAXY

One of the more notable constellations of the Zodiac is Sagittarius, the Archer. It contains no first-magnitude star, but quite a number of stars around the second and third magnitudes. There is no really distinctive pattern, though some people claim that the shape is a little like that of a teapot.

Sagittarius is most significant because it contains the lovely star-clouds which hide our view of the centre of the Galaxy. As has been known for many years, the Galaxy is a flattened system; when we look along the main thickness we see many stars in almost the same direction, which explains the Milky Way effect. But the centre of the system is hopelessly veiled. We are trying to look through what may

The Trifid Nebula, photographed by E.M. Lindsay from the Boyden Observatory, South Africa.

be termed a cosmic fog. Fortunately there are other methods at our disposal. Radio waves can pass through the star-clouds, and so can some of the infra-red radiations. The IRAS or Infra-Red Astronomical Satellite, which was launched in January 1983 and remained active for most of the rest of the year, has proved to be particularly informative.

We have been finding out some very curious things. There are features in the region of the centre of the Galaxy which are expanding outwards, but infra-red studies have also shown that there is a condensed cluster of stars in the central area. Apparently there are about a million of them; we cannot actually see them, but we know that they are there. If we lived on a planet circling a star in this region, there would be no darkness, and the night sky would be ablaze.

There is one odd factor. The total mass of the stars in the cluster does not seem to be great enough to prevent the cluster from dispersing. There ought to be at least five times as much. The missing mass must be somewhere or other—and, inevitably, there are suggestions that it is concentrated in a black hole.

Black holes, as we know, are bizarre by any standards. They are due to old, collapsed stars pulling so strongly that nothing can escape from them—not even light. So the old star is surrounded by a region which is to all intents and purposes cut off from the outer universe. If there really is a black hole in the galactic centre, it may have a mass about five million times that of the Sun, and it will be in the nature of a cosmic plughole which is sucking material in from all around. Other galaxies, too, are believed to have central black holes, and the same may be true of some of the globular star-clusters.

All this is highly controversial, but certainly there are strange phenomena associated with the core of the Galaxy. Whether or not it means that a black hole is present there remains to be decided, but in any case it is an intriguing possibility.

116 THE PAST AND FUTURE SKY

In ancient times the stars were usually called 'Fixed Stars', to distinguish them from the wandering stars or planets. Of course, the constellations do seem to remain unaltered over long periods, because the stars are so far away; but over a sufficient length of time, the small 'proper motions' of the stars become evident. It was Edmond Halley who first realised that several of the most brilliant stars in the sky had shifted slightly since Greek times, compared with the more remote stars in the background. The greatest proper motion known is that of Barnard's Star, a faint red dwarf a mere six light-years away. In 180 years, Barnard's Star moves by a distance equal to the apparent diameter of the full moon.

But looking back and forward, it is evident that the constellations we know will become unrecognisable. For example, in the Great Bear or Plough, two stars (Alkaid and Dubhe) are travelling in a direction opposite to that of the other five. The apparent magnitudes of the stars will alter, too. Which would have been supreme a million years ago?

Calculations show that in one million years before Christ the brightest star in the whole sky was Saiph or Kappa Orionis, which would have been of magnitude − 4.3—equal to Venus. Of course, it would not have been in the same position as it is today; it would have been in the far north of the sky. Today it has faded to below magnitude 2. A million years ago the second brightest star was Zeta Sagittarii (− 2.7, as brilliant as Jupiter); then followed Alpha Columbæ (− 1.5) and then Canopus (− 0.8). At the present moment both Zeta Sagittarii and Alpha Columbæ have faded to magnitude 2.6.

What, then, about the sky a million years hence? We find that the leader will be Eltamin or Gamma Draconis, which is in the Dragon's head today but which will have shifted to the area now occupied by Ophiuchus, the Serpent-bearer. Eltamin will have brightened up to magnitude − 2.2. Then will come Delta Scuti (− 1.3), Menkarlina or Beta Aurigæ (− 0.7), Gamma Scuti (also − 0.7) and Canopus (− 0.6). At present Eltamin is of magnitude 2.2 and Menkarlina 1.9, but Delta and Gamma Scuti are much fainter: each is 4.7, though Delta is slightly variable. Obviously, then, both these stars will have come much closer to us. Each is of the order of 300 times as luminous as the Sun.

The most notable fact about these alterations is that only Canopus retains its position in the 'top five'. This is because it is so powerful and so far away that its slight change in distance will have little effect on its brilliancy as seen from Earth. Not so with its only superior in the present sky, Sirius, which was below the second magnitude a million years ago and will again be so in a million years' time.

117 JUPITER WITHOUT ITS MOONS

Any small telescope will show the four bright satellites of Jupiter: Io, Europa, Ganymede and Callisto. Good binoculars will also show them, and a few keen-sighted people can see them with the naked eye. But for the fact that they are so drowned by the glare from Jupiter, three of them at least would be easy naked-eye objects.

Very occasionally, however, Jupiter appears moonless. The bright satellites undergo various phenomena, and on rare occasions all four are affected at the same time. First, a satellite may pass in transit across the face of Jupiter, so that it is seen against the disk of the Giant Planet. Generally the outer satellites, Ganymede and Callisto, look darkish, while the more reflective inner satellites, Europa and Io, look bright; but it is not easy to see any of them when they are full on the disk, though the shadows which they cast are more evident. Secondly, a satellite may pass behind Jupiter. It is then hidden or occulted, and is of course out of view. Thirdly, a satellite may pass into the cone of shadow cast by Jupiter, and be eclipsed.

One of the rare occasions when Jupiter appeared moonless was on April 9 1980. For a period following 13:13 hours GMT, Io was in transit in front of Jupiter, Ganymede and Europa were occulted behind Jupiter, and Callisto was eclipsed by Jupiter's shadow. At 14:15 hours Ganymede reappeared from occultation, and for 19 minutes it was the only satellite visible. At 14:34 hours Ganymede was eclipsed by Jupiter's shadow, and Jupiter again seemed moonless until 15:28 hours, when Io emerged from transit.

There are, of course, phenomena of added interest. Io and Europa generally look bright soon after entering transit, but are usually lost as they approach the central meridian, to reappear later just before the transit ends. The shadow transits are easy to follow, and so are the eclipses. In fact, it was by timing the eclipses of Jupiter's satellites that the Danish astronomer Ole Rømer first worked out the speed of light. When Jupiter was at its furthest from Earth, the eclipses were late, because of the extra time needed to cross the space between the two planets.

Incidentally, it seems that the 'moonless' periods were worked out around 1910 by an Italian amateur astronomer named Enzo Mora. He made extensive calculations of the movements of the satellites, and was remarkably accurate. But how did he do it?—because really precise tables of the motions of the satellites were not published until some time later. Mora wrote in French from an Italian address in a German periodical, and very little seems to be known about him, but at least the predictions which he gave turned out to be remarkably correct.

118   HOW TO WEIGH A BLACK HOLE

Black holes are undoubtedly the weirdest objects in the universe. They are not true 'holes', but areas round old, collapsed stars which are pulling so powerfully that not even light can escape from them. The old stars are, therefore, surrounded by a region which is to all intents and purposes cut off from the rest of the universe.

We cannot see black holes; we can only observe their effects upon objects which we can detect. And by now it is widely believed that black holes lie in the centres not only of some galaxies, but also in the remarkable, super-luminous quasars, some of which cannot be far from the edge of the observable universe.

How can one 'weigh' a black hole? An ingenious method has recently been worked out by a team of British astronomers, who have been concentrating upon a galaxy known only by its catalogue number of NGC 4151. (It lies in the constellation of Canes Venatici, the Hunting Dogs, near the Great Bear; I can see it with the 15-in reflector in my observatory at Selsey, but not easily.) NGC 4151 varies in brightness. If there is an outburst in the centre of the system, it takes some time to become evident in the outer regions. There is an inevitable delay; and this means that it is possible to measure the distance between the centre of the galaxy and its edge. It is also possible to find out how fast the outer regions are moving round the centre; like all galaxies of its type, NGC 4151 is rotating.

The method may be compared with that used to measure the mass of the Sun. Planets move round the Sun, and their distances and velocities show how powerful the Sun's gravity really is. We can make the same calculations for the different parts of NGC 4151 as they rotate round the core. And if we assume that the core is occupied by a black hole, we can find that the mass of the black hole is between 50 and 100 million times that of the Sun.

Astounding though it may seem, this result has not really surprised astronomers. NGC 4151 is what is termed a 'Seyfert' galaxy (after Carl Seyfert, who first described systems of this type). There is a condensed nucleus, and only weak spiral arms. It now looks as though the inner part of the nucleus is a black hole; it may be described as a 'mini-quasar'.

Researches of this kind are still in an early stage, and there are bound to be uncertainties; but it does look as though we have been able to give at least a reasonable value for the mass of an object which we will never be able to see.

119 GEMINGA

X-ray astronomy began in 1963, when it became possible to send instruments up by rocket above the top of the atmosphere. Gamma-rays are even shorter than X-rays, and are more difficult to study. Nevertheless, gamma-ray sources have been identified. One of them is the Crab Nebula, the remnant of the supernova which flared up in 1054 (or, rather, was seen in 1054; the Crab is several thousands of light-years away). Another is in the constellation of Gemini, and is usually called Geminga—*Gemini ga*mma-ray source. It was identified by an artificial satellite launched in 1972, and further studies of it were made with later satellites.

At gamma-ray wavelengths, Geminga is one of the strongest sources in the sky; but nothing prominent can be seen visually in its position. Searches were then made at X-ray wavelengths, and several candidates were found, one of which was presumably Geminga. Optical searches were also carried out, and it now seems at least possible that Geminga can be identified with a very faint starlike point of magnitude 23—not so far above the limit of detection even with our largest optical telescopes.

The X-ray spectrum indicates that the distance is about 600 light-years, though obviously there is a wide margin of uncertainty even if the identification is correct. The gamma-ray spectrum is not very unlike that of another powerful but known source, lying in the southern constellation of Vela. But the Vela source is a pulsar, and has been positively identified both by its radio signals and its optical flashing. Geminga emits no radio waves at all.

This leads on to an intriguing possibility. Pulsars are neutron stars, with strong magnetic fields, and send out their pulsed radiation as they spin rapidly round; also, they are gradually slowing down, and although the rate of decrease is almost

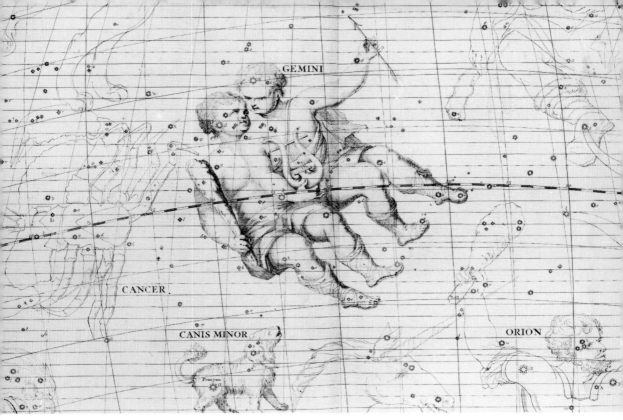

Gemini, the Twins, from Flamsteed's Celestial Atlas, *1729 edition.*

unbelievably small there will come a time when the pulses are no longer detectable. Can Geminga, then, be a pulsar which has stopped pulsing?

This seems to be a favoured theory at the moment, but we cannot come to a definite decision until we are quite sure that we are studying the true source. For the moment, Geminga is guarding its secrets well.

120   WHERE CAN YOU SEE THE SOUTHERN CROSS?

It is often said that the Southern Cross, the most famous of all the constellations to Australians or New Zealanders, is never visible from the northern hemisphere of the Earth. This is not correct. To work out the 'limit of visibility', we have only to do a little simple arithmetic.

A star's declination is its angular distance from the equator of the sky (just as the longitude of any point on the Earth's surface is its angular distance from the terrestrial equator). The southernmost star of the Cross, Acrux or Alpha Crucis, has a declination of $-63°$, or 63 degrees south (for the sake of clarity, let us deal in whole numbers). To find the limiting latitude of visibility, we have only to take 63 away from 90. This gives 27; so in theory Acrux rises from any point on the Earth south of latitude 27°N. In practice it is not likely to be seen below an altitude of about 3 degrees, so we can take latitude 24°N as our limit.

The whole of Europe lies north of this latitude, and so the Cross cannot be seen.

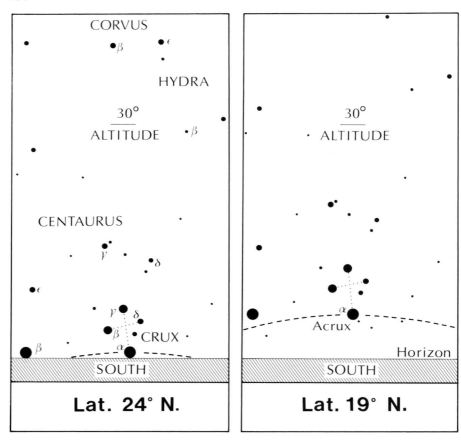

Latitude limits for visibility of the Southern Cross.

What about the Bahamas? Their latitude is about 24°N, and so Acrux could just about be seen; the other Cross stars would be slightly higher. Hong Kong (latitude 22°N) will do; the Cross will rise briefly. From Hawaii, where major telescopes are situated, the latitude is 20°N, and the Cross can certainly be made out; from Mexico City (19°N) it is easy. So to see this glorious constellation there is no need to cross the Earth's equator.

The four main stars in the pattern are Alpha Crucis (Acrux), Beta Crucis, Gamma Crucis and Delta Crucis—the last three have no official proper names, though Beta is sometimes called Mimosa. Let us consider their apparent magnitudes, their absolute magnitudes and their distances. Apparent magnitude, of course, is a measure of how bright a star looks; absolute magnitude is a measure of its luminosity, and is defined as the apparent magnitude that the star would have if it were seen from a standard distance of 32.6 light-years. To complicate matters somewhat, Acrux is a splendid binary; the two components are easily split with a small telescope (there is also a third, unconnected star in the field of view). The results are interesting:

Star	Distance in light-years	Magnitude Apparent	Absolute	Luminosity (Sun = 1)
Acrux	360	0.8	− 3.9 and	3200 and
			− 3.4	2000
Beta Crucis	415	1.2	− 5.0	8200
Gamma Crucis	45	1.6	− 0.5	160
Delta Crucis	257	2.8	− 3.0	1300

So from our standard distance Beta, the most powerful of the four, would shine more brilliantly than Venus does to us. Gamma, the orange member of the group, is the least luminous; Delta looks fainter than Gamma only because it is further away. In astronomy, appearances can often be misleading.

121   THE RED-HOT ASTEROID

On October 21 1983 Simon Green, an astronomer working at Leicester University, was studying some results from IRAS, the Infra-Red Astronomical Satellite, when he made a startling discovery. He detected a rapidly-moving object which had been observed by the telescope in seven consecutive orbits of the Earth, spaced 1 hour 40 minutes apart. What could it be? The news was immediately sent to observatories all round the world, and the object was located and photographed both at the Palomar Observatory in California and the Lowell Observatory at Flagstaff in Arizona. It was very faint—about magnitude 15—and it was presumably an asteroid.

Brian Marsden, of the International Astronomical Union, calculated an orbit and reported that the asteroid was very unusual indeed. It was a mere 1.2 miles in diameter, and when discovered it had been about 19 million miles from the Earth, so that it belonged to the Apollo class—that is to say, it had an orbit which took it closer to the Sun than the path of the Earth. Its orbit was inclined to ours at about 20 degrees, and the revolution period was one-and-a-half years. But—and this was the remarkable fact—its eccentric path took it to a mere nine million miles from the Sun.

No other asteroid was known to approach anything like so close as that. Indeed, only one—Icarus—was known to go in closer than the orbit of Mercury, and the minimum distance between Icarus and the Sun is always greater than 17 million miles. This meant that the new asteroid—given the temporary designation of 1983 TB—was the most extreme member of the whole group. When near perihelion it must be red-hot, though at its greatest distance from the Sun it swings out beyond Mars into the main asteroid zone, and must then be bitterly cold. Without doubt it has the most unpleasant climate in the whole of the Solar System.

Perhaps significantly, its orbit is almost the same as that of the annual Geminid meteor shower, seen every December. It has been suggested that the asteroid is simply a dead comet which has shed its material to produce the Geminids and is now simply an inert remnant of its former self. This may or may not be so, but in any case this peculiar little body will be closely followed. So far as we know, it is unique. One day it may be surveyed by an unmanned rocket probe, but I hardly think that even the most courageous astronaut will feel inclined to visit it.

122 SOLAR ECLIPSES FROM OTHER WORLDS

It is sheer luck that in our skies the Sun and the Moon appear of the same size, so that the Moon can just cover up the face of the Sun during a total eclipse. Other planets, too, have satellites; what then are the chances of seeing total eclipses from there?

No chance at all from Mars, because both the satellites, Phobos and Deimos, are much too small. To us, the Moon's apparent diameter is half a degree. From Mars, Phobos would have a maximum apparent diameter of about 12 minutes of arc, Deimos only two minutes, while the diameter of the Sun would be around 21 minutes. Instead of causing eclipses, therefore, both satellites would appear in transit across the solar disk. Phobos would transit 1,300 times in a Martian year, taking 19 seconds of time to pass from one limb to the other; Deimos would transit 130 times, taking 1 minute 48 seconds each time (these figures are, of course, means; they can vary to some extent either way).

From Jupiter the apparent diameter of the Sun would be six minutes of arc, and five of its satellites could produce total eclipses: Amalthea (apparent diameter 7′24″), Io (35′40″), Europa (17′30″), Ganymede (18′6″) and Callisto (9′30″). From Saturn, where the diameter of the Sun is reduced to 3′22″ on average, total eclipses could be produced by all the major satellites out as far as Titan. To a Saturnian, Tethys would actually seem the largest, with a diameter of 17′36″, followed closely by Titan at 17′10″. From Uranus the diameter of the Sun is a mere 1′41″, so that all the five known satellites would cover it; their diameters range from 30′54″ for Ariel down to 9′48″ for Oberon. And with Neptune, where the Sun has shrunk to 1′4″, Triton would be more than adequate, since

Below left *Dione, one of Saturn's satellite's, seen from Voyager 1 from 500,000 miles on November 12 1980. Note the 'frosty' wisps. From Saturn, Dione could easily cover the sun.*

Below right *Uranus, photographed on March 1 1948, showing its five satellites, Miranda, Ariel, Umbriel, Titania and Oberon.*

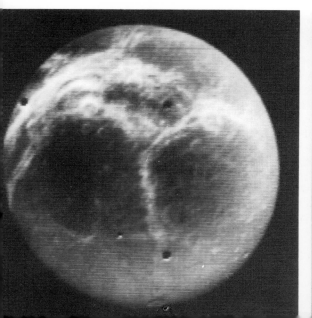
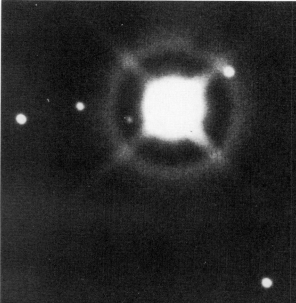

This montage of images of the Saturnian system was prepared from an assemblage of images taken by Voyager 1. This artist's view shows Dione in the foreground, Saturn rising behind, Tethys and Mimas fading into the distance to the right, Enceladus and Rhea off Saturn's rings to the left and Titan in its distant orbit at the top.

from the Neptunian surface it would subtend an angular diameter of just over a degree—larger than our Moon appears to us.

All the same, the Earth-Moon pair provides the only example of a satellite which is virtually equal in apparent diameter to that of the Sun. We are singularly fortunate. But for this coincidence, it would have taken us a long time to learn much about the Sun's outer atmosphere.

123   ALIEN LIFE?

Some time ago I was giving a talk to a scientific society in London, dealing with the possibility of life elsewhere in the universe. I said that there seemed no chance of finding any intelligent being in the Solar System, except (possibly) on the Earth, simply because conditions on the other planets—even Mars—were unsuitable. I was then challenged by a member of the audience, who asked what was a perfectly reasonable question. Simply because *we* cannot live on Mars, is there any objection to picturing a completely different life-form which could manage very well in a cold climate and with very little air?

The point here is that so far as we know, all living material is based upon one atom: carbon. Only carbon atoms have the power to join with other atoms to make up the complicated molecules that are needed for life. Therefore, it is logical to assume that life, wherever it may be found, is carbon-based; and this means that conditions will have to be at least roughly similar to those on Earth, with a reasonably dense atmosphere, an equable temperature and sufficient water.

There is one loophole. The atom of silicon also has some ability to form complicated molecules. However, it is not nearly so efficient as carbon, and there is no evidence of silicon-based life anywhere.

At the moment this is about as far as we can go, but there have nevertheless been some rather bizarre suggestions which cannot be dismissed out of hand. For instance, consider Jupiter. As we know, the outer clouds are very cold, but lower down the temperature rises, and the planet's core is extremely hot. Moreover, much of Jupiter is liquid. The Russian astronomer Iosif Shklovsky (referred to in article 94 above in connection with the satellites of Mars) made the following interesting comment in 1966:

'Jupiter is in fact an immense planetary laboratory. For example, we can imagine organisms in the form of ballasted gas-bags, floating from level to level in the Jovian atmosphere, and incorporating pre-formed organic matter, much like plankton-eating whales of the terrestrial oceans. . . . When the preliminary detailed reconnaisance of our Solar System is completed a century hence, it may well turn out that the greatest surprises and the most striking advances for biology attended the exploration of Jupiter.'

It is certainly an intriguing thought, though life inside Jupiter would be rather hazardous—the 'gas-bags' would have to keep to fairly strict limits, avoiding both the intense cold of the upper regions and the fierce heat lower down. I admit that I am profoundly sceptical. But—well, who knows?

124   THE FURTHEST QUASAR

Every few months there seems to come an exciting new discovery. Such was the case in the first part of 1982. A new quasar was found—a very special one, because it is the most remote and possibly the most luminous object ever detected.

Quasars were first identified in 1963, and have provided puzzle after puzzle ever since. Superficially they look rather like stars, but they are very different; they are comparatively small, and super-luminous. One quasar may equal the output of a hundred whole galaxies—and remember that a galaxy such as our own contains 100,000 million suns. Some quasars, though not all, are strong radio sources, and it seems to be a general rule that the remoter quasars are powerful in the radio range. Quasars are now believed to be the nuclei of very active galaxies, possibly containing central black holes. At any rate, they are bizarre by any standards.

In 1972 a quasar catalogued as OQ 172 was found, and seemed to be more remote than any previously discovered. Yet it was not among the faintest quasars known, and had it lain still further away it could still have been detected. Why were there apparently no quasars more remote than OQ 172? Had we reached the limit of observation?

Most astronomers doubted this. There must be quasars at greater distances. Using the 210-ft 'dish' radio telescope at Parkes in Australia—still one of the largest and best radio telescopes in the world, even though it was completed more than two decades ago—Dr Alan Wright and his team searched away. Finally they

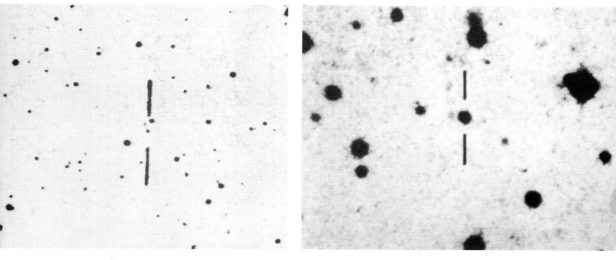

Above left *Quasar OQ 172, former holder of the distance record.*

Above right *PKS 2000-330, the most remote quasar know at present. It is the dot between the two vertical lines. (This is a negative, with white and black reversed.)*

found what they had hoped. They sent the position of the radio source to the Siding Spring Observatory, and astronomers there identified the radio source with an optical object. It looks like a dim blur, but it is immensely significant.

How far away is it? According to the red shift of the lines in its spectrum, it appears to lie at least 13,000 million light-years from us, in which case we are seeing it as it used to be 13,000 million years ago, when the universe was still young and the Earth lay in the far future. Also, the quasar is racing away at over 91 per cent of the velocity of light, so that it is much more remote now than it was when you started reading this page. Its catalogue number is PKS 2000-330.

Whether we have now reached the limit remains to be seen. I suspect that we have not, and that in the course of time the record of PKS 2000-330 will be broken. It is a rule that increased velocity of recession is linked with increased distance; the further away an object such as a quasar is, the faster it is going. If so, then at 15,000 million light-years we may reach a point at which a quasar is receding at the full velocity of light, and will be beyond our range. This would be the boundary of the observable universe. PKS 2000-330 is not so remote as that— but it seems to be not so very far from it.

125 TO CATCH A COMET

As I write these words (February 1984) one of the most remarkable of all space probes is being built. It is named Giotto, and is being constructed at the British Aerospace factory at Filton, near Bristol; I saw it yesterday. On March 13-14 1986, all being well, it will meet Halley's Comet in headlong collision.

Myself, with the half-built Giotto, at Bristol on January 19 1984.

At the moment we have to admit that we know very little about the heart of a comet—its nucleus, where almost all the mass is concentrated. As a comet approaches the Sun, its inner ices start to vapourise, and the nucleus is hidden behind an opaque veil which no telescope can pierce. The only real way to find out is to send a space-craft, preferably to a large comet rather than one of the smaller, faint periodical comets which have been back to the neighbourhood of the Sun so often that they have lost most of their volatiles. And the only major comet which comes back regularly, and which we can predict, is Halley's.

It came to perihelion in 1910, and was last seen in 1911, after which it was beyond our ken until it was picked up again in October 1982 by observers using the 200-in Palomar reflector. It next reaches perihelion on February 9 1986 (though, unfortunately, conditions are about as unfavourable as they can possibly be, and Halley will not be nearly so brilliant as it has been in the past). However, it provides our only opportunity until the return of 2061, which is rather a long time to wait. The Americans cancelled their comet probe on the grounds of expense, but the Russians are sending two and the Japanese are also in the field. European hopes rest with Giotto, named after the painter who used the comet to represent the Star of Bethlehem in his picture *The Adoration of the Magi.*

Giotto, due to be launched in mid-1985 from the French base at Kourou in Guiana, is not a large probe. It has been kept deliberately small, because during the last, vital stages of its journey it is bound to be peppered by cometary material, and the smaller the target the better. Things are complicated by the fact that Halley has a retrograde orbit, so that the comet and the probe will meet head-on at the tremendous relative speed of 150,000 mph. Giotto has been provided with a dust-shield, but not even the most optimistic planners really expect it to go right through the comet and emerge unscathed. It will enter the cometary 'atmosphere' and pass within 310 miles of the nucleus. The critical encounter time will last for four hours, during which the results will be transmitted back to Earth, but the most important information will be obtained during a period of only about two *minutes!*

Will Giotto succeed—or will the Russians or the Japanese carry through their projects faultlessly? Time will tell. By March 15 1986 we should, with luck, know a great deal more about comets than we do now. But luck is, frankly, needed. There can be no rehearsal; there is only one Giotto, and there can be no second chance.

126   THE MARTIANS OF AD 3000

Long ago, during my boyhood (and that goes back to the 1930s), fictional BEMS, or Bug-Eyed Monsters, were all the rage. Alien planets were populated with creatures of all kinds, some of which lived in seas of liquid methane while others had six or seven heads, innumerable tentacles, metallic skins or long, scaly tails. Many were telepathic, and almost all of them were decidedly peevish by nature.

Today the BEMs have largely vanished from literature, and our ideas about life elsewhere have become much more positive. In particular, we can rule out intelligence anywhere in the Solar System except (possibly!) on the Earth. Mars, the most plausible candidate, has proved to be a world of mountains, valleys, volcanoes and craters, but with a hopelessly thin atmosphere made up chiefly of carbon dioxide. The Viking probes found no sign of life; if there is any Martian life at all it must be very lowly.

In the future, things should be different. Barren though it may be, Mars is much less unlike the Earth than any other member of the Sun's family, and there is plenty of ice, so that there should be no shortage of H_2O. It is bound to be our first space-target after the Moon, but the problems are much greater. Mars never comes much within 35 million miles of us, and it does not keep company with us as we travel round the Sun. Rockets of 1984 type must take months to get there, and though the time of travel will eventually be reduced we must admit that journeys to Mars will always be fairly lengthy. Moreover, there must be a period of waiting on the planet before it and Earth are suitably placed for the return journey. This means that even the pioneer explorers will have to set up a Martian Base. Design studies have already been made. I do not propose to discuss them here, because I want to look further ahead at least a hundred years—let us say to the year AD 3000. By then there should be permanent colonies on Mars. There will be men, women and—inevitably—children.

Mars, remember, has a surface gravity only 0.38 of that of the Earth. There is no reason to doubt that *homo sapiens* can adapt to such conditions; but Martian babies would grow up under 0.38 g, and presumably their muscles would develop

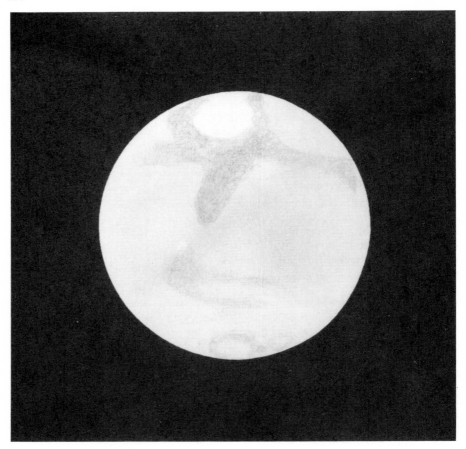

Mars, as I drew it from a 10-in refractor.

accordingly. What, then, would happen if a Martian boy or girl were taken to Earth? There would be a feeling of terrible heaviness; moving around would be an effort. Could 'Martian' muscles cope?

The answer may well be 'No'. In that case, we will have a situation in which Earthmen can go to Mars and live there, but Martians can never come to Earth. Their nearest approach to the world of their ancestors would be either the Moon or else an orbiting space-station. They could look down on the green fields, the forests and the oceans and lakes, knowing that to go there would be fatal. They would be fully entitled to regard the Earth as a planet of death.

Of course, this may be an extreme viewpoint, because we do not yet know nearly enough about the effects of prolonged exposure to reduced or zero gravity. But it is a situation which could arise, and it is possible that men who emigrate to Mars will do so with the knowledge that they will never be able to return. Yet it does seem overwhelmingly likely that unless civilisation is wiped out by a third world war, Mars will be colonised during the next century. The Solar System will have two inhabited worlds instead of one; there will be Martians at last.

GLOSSARY

Aphelion The furthest distance of a planet or other body from the Sun in its orbit.

Apparent magnitude The apparent brightness of a celestial body. The lower the magnitude, the brighter the object. Thus a star of magnitude 1 is brighter than a star of magnitude 2, 2 is brighter than 3 and so on; stars of magnitude 6 are normally the faintest visible with the naked eye, but modern instruments can record objects down to magnitude 26.

Asteroids (or *minor planets*) Small planets moving round the Sun; only Ceres is as much as 500 miles in diameter. Most of the asteroids keep to the region between the orbits of Mars and Jupiter, though some swing well away from the main swarm.

Astronomical unit The distance between the Earth and the Sun; in round numbers 93 million miles.

Binary star A star system made up of two stars, genuinely associated, and moving round their common centre of gravity. The revolution periods range from a few hours to millions of years.

Black hole A region round an old, collapsed, very massive star which is now pulling so strongly that not even light can escape from it.

Bode's Law A mathematical relationship between the distances of the planets from the Sun. It breaks down for Neptune, and is probably nothing more than coincidence.

Cepheid A short-period variable star. The period is linked with the real luminosity, so that once the period has been found the luminosity—and hence the distance—may be calculated.

Corona The outermost part of the Sun's atmosphere, made up of very tenuous gas.

Cosmic rays Not true rays, but high-velocity atomic particles coming in from all directions in space.

Declination The angular distance of a celestial body from the equator of the sky. Polaris has a declination of over 89° N, so that it is within one degree of the north celestial pole; there is no bright south polar star.

Doppler effect The apparent change in the wavelength of light from a luminous body as it approaches or recedes. With an approaching body the light is 'too blue'; with a receding body, 'too red'. The changes show up in the shifts of the dark lines in the spectra of stars and galaxies.

Earthshine The faint luminosity of the night side of the Moon, due to light reflected on to the Moon from the Earth.

Eclipse, lunar The passage of the Moon through the shadow cast by the Earth. Lunar eclipses may be either total or partial.

Eclipse, solar The blotting-out of the Sun by the Moon, when the Moon passes between the Sun and the Earth. Solar eclipses may be total (all the Sun covered), partial, or annular (when the Moon is at its greatest distance from the Earth and appears smaller than the Sun, so that a ring of the Sun is left showing round the dark disk of the Moon).

Escape velocity The minimum velocity which an object must be given to escape from the pull of a planet or other body. The escape velocity of the Earth is seven miles per second.

Galaxies Systems made up of stars, nebulæ and interstellar matter. Many, though by no means all, are spiral in form.

Galaxy, the The Galaxy of which our Sun is a member; it contains about 100,000 million stars, and is a loose spiral.

Gamma-rays Radiations of extremely short wavelength.

Gibbous phase The phase of the Moon or a planet when between half and full.

Inferior planets Mercury and Venus, which are closer to the Sun than we are. They are the only planets showing a complete cycle of phases, from new to full.

Libration The apparent 'tilting' of the Moon as seen from the Earth, so that at various times we can examine 59 per cent of the total surface—though, of course, no more than 50 per cent at any one time.

Light-year The distance travelled by light in a year: almost six million million miles.

Meteor A small particle, usually smaller than a pin's head, which dashes into the upper air and destroys itself in the streak of luminosity which we call a shooting-star.

Meteorite A larger body which lands on Earth without being burned away. Meteorites are not large meteors, and are more nearly related to the asteroids. In space, all these stray bodies are known collectively as *Meteoroids.*

Nebula A cloud of gas and dust in space, inside which new stars are being formed from the nebular material.

Neutrino A fundamental particle with no electrical charge, and no mass (or at least, very little).

Neutron A fundamental particle with no electrical charge, but a mass almost equal to that of a proton (qv).

Nova A star which flares up suddenly to many times its normal brilliancy, remaining bright for a while before fading back to its former obscurity. Novæ are believed to be binary systems.

Occultation The covering-up of one celestial body by another. An eclipse of the Sun should really be called an occultation of the Sun by the Moon.

Opposition The position of a planet when exactly opposite to the Sun in the sky, so that it is best placed for observation.

Orbit The path of a celestial body in space.

Perihelion The position in orbit of a body or other object when closest to the Sun.

Perturbations Disturbances in the orbit of a celestial body, due to the gravitational pulls of other bodies.

Planetary nebula A small, hot, dense star surrounded by a shell of gas. Planetary nebulæ are not planets, and are not true nebulæ, so that the name is decidedly inappropriate.

Position angle The apparent direction of one celestial object with reference to another. With double stars, the position is taken as being the direction of the fainter component referred to the brighter.

Prominences Masses of glowing hydrogen above the surface of the Sun. With the naked eye they can be seen only during total eclipses, but spectroscopic equipment means that they can be studied at any time.

Proper motion, stellar The individual movement of a star against the background of more distant stars.

Proton A fundamental particle with unit positive charge. Round it move *electrons*, which have unit negative charge though a much lower mass.

Quasar A very remote, super-luminous system. Quasars are now believed to be the nuclei of very active galaxies, and are probably 'powered' by black holes.

Radiant The point in the sky from which the meteors of any particular shower seem to come. Thus the August Perseids radiate from a position in the constellation of Perseus.

Retrograde motion Orbital or rotational movement in the sense opposite to that of the Earth.

Right ascension The angular distance of a body from the spring equinox.

Scintillation Twinkling of a star (or planet). It is due entirely to the unsteady atmosphere of the Earth.

Selenography The study of the surface of the Moon.

Seyfert galaxies Galaxies with small, bright nuclei and weak spiral arms. Many of them are strong radio sources.

Sidereal period The revolution period of a planet or other body round the Sun, or of a satellite round a planet. The Earth's sidereal period is, of course, one year (more accurately, 365¼ days).

Solar wind A flow of atomic particles streaming out from the Sun in all directions.

Spectroscope An instrument for splitting up light, thus giving information about the composition of the light-source.

Superior planets Planets lying beyond the orbit of the Earth in the Sun's family or *Solar System*: that is to say, all the main planets apart from Mercury and Venus.

Supernova A violent stellar outburst, in which a formerly faint star flares up and may become at least 15 million times as luminous as the Sun for a brief period.

Terminator The boundary between the sunlit and night hemispheres of the Moon or a planet.

Transit (of Mercury or Venus) The apparent crossing of the Sun's disk by an inferior planet.

Van Allen zones Zones of charged particles around the Earth.

Variable stars Stars which change in brilliancy over comparatively short periods. They are of various types, some regular, others irregular.

White dwarf A very small, dense star which has used up all its nuclear energy.

Zenith The observer's overhead point (altitude 90 degrees).

Zodiac A belt stretching round the sky, eight degrees to either side of the *ecliptic* (the Sun's apparent yearly path round the sky), in which the Sun, Moon and bright planets are always to be found.

Zodiacal Light A cone of light rising from the horizon and stretching along the ecliptic. It is visible only when the Sun is a little way below the horizon, and is due to thinly-spread interplanetary matter near the main plane of the Solar System.

INDEX